Contents

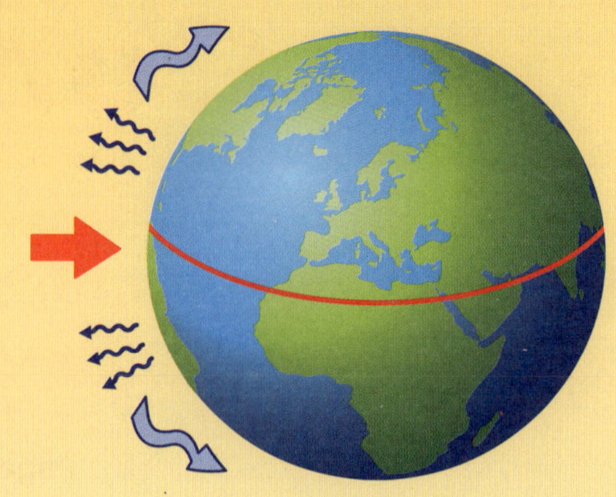

TARGETING SCIENCE YEAR 4 © PASCAL PRESS ISBN: 978125726534

Targeting Science Year 4

ISBN: 9781925726534

Published by Pascal Press
PO Box 250
Glebe NSW 2037
www.pascalpress.com.au
contact@pascalpress.com.au

This edition of *Skill Sharpeners: Science* is published by arrangement with Evan-Moor Corporation, USA.

For sale in Australia and New Zealand.

Authors: Guadalupe Lopez, Lisa Vitarisi Mathews
Publisher: Lynn Dickinson
Cover design: Janice Bowles
Illustrator: Paul Lennon, *www.dreamstime.com.au*
Editor Australian edition: Stella Tarakson
Typesetter: Stacey Grainger

Printed in China by 1010 International Ltd.

Introduction

Welcome to your Year 4 *Targeting Science* activity book! It is packed with interesting and exciting activities to help you understand and enjoy science at home or school.

Targeting Science has been written to support the Australian Primary Science Curriculum Version 9.0 and is divided between:

- Biological Sciences
- Earth & Space Sciences
- Physical Sciences
- Chemical Sciences

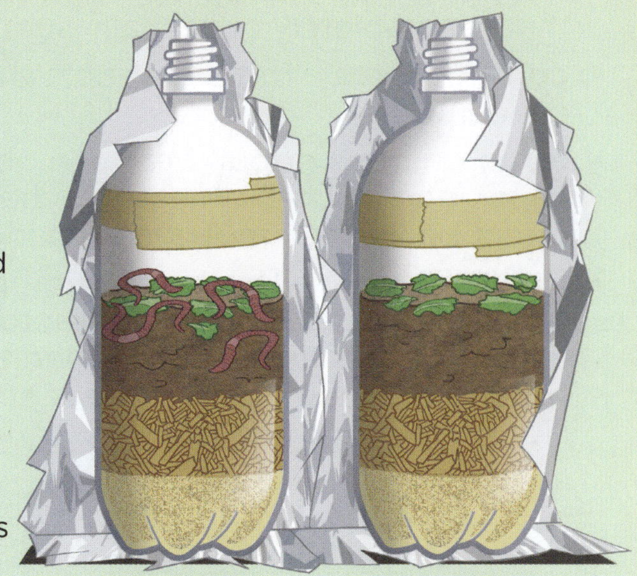

The Science Understanding and Inquiry Skills elements (including Science as a Human Endeavour) are embedded.

Hands-on activities provide opportunities to bring the science concepts to life. The exercises develop process skills such as observing, collecting, recording and organising information and making models of scientific events that happen in the natural world. All activities use easily sourced, inexpensive items.

Topics are introduced with explanations plus images to provide background information. Some lessons include a QR code where you can access a video for further explanations. You will be challenged to match, sort, label, sequence, analyse and answer questions with regular vocabulary practice puzzles throughout the book to help familiarise you with scientific terms. See the Glossary on the next page for some terms you may not already know. Answers are included at the end of the book.

FREE Teaching Guide.

This QR code links to a downloadable PDF of a Teaching Guide to support the material in this student workbook. The guide contains:

- Teaching plans and checklists.
- Graphic organisers.
- Material request forms (to send home for parents).
- Background information about each major topic as well as extension activities.

Use this QR code to access the FREE Teaching Guide.

Front cover – have you looked at the front cover? It depicts what life was like before a significant scientific discovery that led to the development of new technology. What is the difference between Science and Technology? The following quote sums it up nicely.

Science is the process of acquiring knowledge of natural phenomenon along with various reasons. Technology is the application of Science to the solution of problems. (Sismondo, 2018)

Glossary

- **buoyant force** - an upward force exerted on an object that is completely or partially submerged in liquid.
- **contact force** – force that acts through direct physical contact.
- **force** - a push or pull between objects, which may cause one or both objects to change speed and/or direction of their motion or change their shape.
- **force diagram** – Here is an example of a 'force diagram' where this man on a tractor is trying to pull a log. The force acting between the log and the ground is friction. This force is acting in the opposite direction to the tractor – trying to slow it down.

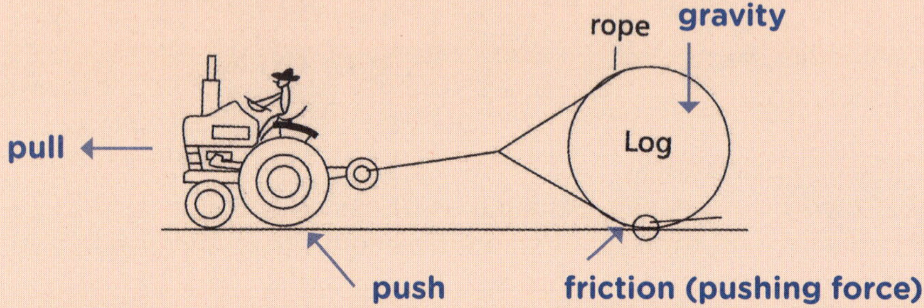

- **friction** – the force that resists movement when one surface slides against another.
- **gravity** – the invisible force of attraction between any two objects. Gravity holds the Universe together, keeps the planets in orbit around the Sun and satellites in their orbits.
- **groundwater** – the water beneath the surface of the ground, which consists largely of surface water that has seeped down, and which eventually drains into rivers, lakes and wetlands.
- **non-contact force** – force applied through invisible forces like gravity or magnetism without any direct physical contact.
- **ocean** - the vast body of salt water which covers almost three-quarters of the Earth's surface.
- **properties** - attributes of an object or material, normally used to describe attributes common to a group.
- **pulling** – a force that changes the direction of an object towards you, is a pull or pulling force.
- **pushing** - a force that changes the direction of an object away from you, is a push or pushing force.
- **single-use plastics** – designed to be used once then disposed of or destroyed.
- **surface water** – water that collects on the surface of the ground.

SAFETY
All of the investigations in this book are designed for kids to do safely in the home or classroom. However, adult supervision is recommended.

TARGETING SCIENCE YEAR 4 © PASCAL PRESS ISBN: 978125726534

Year 4 Australian Science Curriculum Correlations

ACARA Code	Content description	Strand	Sub strand	Pages
AC9S4U01	Explain the roles and interactions of consumers, producers and decomposers within a habitat and how food chains represent feeding relationships	Science Understanding	Biological Sciences	2-26
AC9S4U02	Identify sources of water and describe key processes in the water cycle, including movement of water through the sky, landscape and ocean; precipitation; evaporation; and condensation	Science Understanding	Earth & Space Sciences	27-60
AC9S4U03	Identify how forces can be exerted by one object on another and investigate the effect of frictional, gravitational and magnetic forces on the motion of objects	Science Understanding	Physical Sciences	61-102
AC9S4U04	Examine the properties of natural and made materials including fibres, metals, glass and plastics and consider how these properties influence their use	Science Understanding	Chemical Sciences	103-118
AC9S4H01	Examine how people use data to develop scientific explanations	Science as Human Endeavour	Nature and Development of Science	27, 75, 106, 113
AC9S4H02	Consider how people use scientific explanations to meet a need or solve a problem	Science as Human Endeavour	Use and Influence of Science	21, 66, 74, 79, 81, 88, 94, 103, 106, 107
AC9S4I01	Pose questions to explore observed patterns and relationships and make predictions based on observations	Science Inquiry	Questioning and Predicting	11, 16, 17, 36, 42, 59, 77, 84
AC9S4I02	Use provided scaffolds to plan and conduct investigations to answer questions or test predictions, including identifying the elements of fair tests, and considering the safe use of materials and equipment	Science Inquiry	Planning and Conducting	16, 17, 36, 70, 71, 76, 77, 85, 92, 100, 102, 111
AC9S4I03	Follow procedures to make and record observations, including making formal measurements using familiar scaled instruments and using digital tools as appropriate	Science Inquiry	Planning and Conducting	16, 17, 36, 42, 50, 58, 68, 71, 76, 84, 101, 110, 112
AC9S4I04	Construct and use representations, including tables, simple column graphs and visual or physical models, to organise data and information, show simple relationships and identify patterns	Science Inquiry	Processing, Modelling and Analysing	6, 7, 8, 9, 17, 22, 27, 31, 35, 47, 56, 60, 64, 75, 93, 108, 114, 116
AC9S4I05	Compare findings with those of others, consider if investigations were fair, identify questions for further investigation and draw conclusions	Science Inquiry	Evaluating	17, 18, 75
AC9S4I06	Write and create texts to communicate findings and ideas for identified purposes and audiences, using scientific vocabulary and digital tools as appropriate	Science Inquiry	Communicating	10, 14, 24, 26, 29, 30, 33, 44, 49, 52, 71, 86, 93, 94

First Nations Perspectives on Environment

The water, air, trees, rocks, plants, animals and landforms of Australia are of great importance for First Nations peoples. They consider the land to be a part of them. They believe all things in nature are connected to each other, and that their culture, belief, identity and way of life are all connected.

These cultures are the longest continuing living cultures on record. By working with, instead of against the environment, they have survived for thousands of years. The belief in this connection has been shared between generations of First Nations families and is still practiced today.

Read some of the examples below of how First Nations people have protected their Country throughout history.

Food:	**Fire Management:**
Australian First Nations peoples were hunters and gatherers. Men hunted mainly for larger animals, such as kangaroos, emus, birds, reptiles, and fish. Women and children hunted small animals and collected fruits, honey, insects, eggs, and plants. They only ever took what they needed, and nothing was ever wasted.	Fire management is the use of small, controlled fires to keep trees and shrubs from growing too thick. It encourages new growth. The heat from fires causes some seeds to germinate (sprout). The growth of plants attracts animals to feed in the area and also renews natural resources. Knowing where and when to light these fires is a skill that been passed down over generations.
Sustainable Practises:	**Animal Sanctuaries:**
First Nations peoples who lived in coastal areas carefully managed the mangroves and used them in a sustainable way for thousands of years. For example, they collected crabs only when they were in season. After the yearly harvest, the crabs and mangrove area were left alone to recover until the next crab season. This ensured that the crabs and mangroves were still there in the future.	First Nations peoples used their long-held knowledge to protect parts of the environment that provided sanctuary for certain animal species. Within these areas, no hunting, fishing, burning or gathering was allowed. The sites were refuges to protect a breeding or nesting ground for a particular species, as well as the organisms in that area.

Ecosystems and the Roles of Organisms

Concepts:

Organisms play different roles in their ecosystems.

Organisms in an ecosystem interact with one another through the transfer of energy.

Define It!

consumer: an organism that relies on other plants and animals for food

decomposer: an organism that breaks down dead plant and animal matter

ecosystem: a group of organisms and the environment in which they live

producer: an organism that makes its own food

An ecosystem is made up of a group of organisms that interact with one another and their environment. This natural community can include the entire planet or be contained in a single drop of water. Examples of ecosystems include a tropical rainforest, a pond, a desert, or a garden. Each organism, from the lowly earthworm to the kingly lion, plays an important role.

Organisms in an ecosystem interact with one another through the transfer of energy. Producers (plants) provide energy for the consumers (animals) that eat them. Decomposers (bacteria, fungi, or worms) break down dead plant and animal matter. Then they recycle the organic material for the producers to use again. This transfer of energy sustains all the ecosystems on Earth.

The tropical rainforest ecosystem contains millions of organisms interacting with each other.

Write whether the organism is a *producer*, *consumer*, or *decomposer*.

1. a potted plant ____________________
2. a garden earthworm ____________________
3. a barnyard mouse ____________________
4. a freshwater alligator ____________________

Food Chain

Concepts:

Herbivores, carnivores, and omnivores are consumers.

Organisms in an ecosystem are part of a food chain.

https://clickv.ie/w/jwgx

Use this QR code to access a video on this topic.

Every organism plays a special role in order to keep an ecosystem balanced and healthy. While producers make their own food and decomposers feed on dead matter, consumers must find plants and hunt other animals for food. There are three kinds of consumers: **Herbivores** eat only plants. For example, the grasshopper in the diagram below. **Carnivores** eat only meat. And **omnivores** eat both plants and other animals.

Consumers take the top spots in the **food chain**, with carnivores and omnivores usually ranking highest. This is because carnivores and omnivores are **predators** who kill and eat their **prey**. For example, in a grassland ecosystem, a predator snake might be at the top of a food chain. A grasshopper is eaten by a mouse, which in turn is eaten by the snake. Although the idea of a food chain may seem simple enough, the relationship between predators and prey in an ecosystem can be complex. For example, lions eat crocodiles, but crocodiles have been known to eat lions, too.

Define It!

carnivore: an animal that eats other animals

food chain: a series of organisms in which each member feeds on the one below it

herbivore: an animal that eats only plants

omnivore: an animal that eats both plants and other animals

predator: an animal that kills and eats other animals

prey: an animal hunted or caught for food by another animal

Use the diagram and passage to answer the questions about the food chain. **Hint:** The arrows point from the organisms that get eaten to the organisms that do the eating, because this is the direction the energy is being transferred.

1. Which two animals are predators? ______________________

2. Which two animals are prey? ______________________

3. Which animals are both predator and prey in this diagram? ____________

____

TARGETING SCIENCE YEAR 4 © PASCAL PRESS ISBN: 978125726534

Food Web

Define It!

competition: a struggle between two or more organisms for a limited resource

food web: a network of interconnected food chains

Predators can be at the top of more than one food chain. They may eat everything from hoofed animals that eat grass to small mammals. Those small animals in turn feed on trees, seeds, or insects. In addition, predators are often in competition with other predators who are at the top of their own food chains. These overlapping and connected food chains all make up one giant food web.

The top predators in an ecosystem play an important role of controlling the number of animals below them in the food web. Without predators, other animal populations could greatly increase. And with so many populations competing for limited food and water, these animals could die from starvation. So every organism, no matter where it is in the food web, helps other organisms.

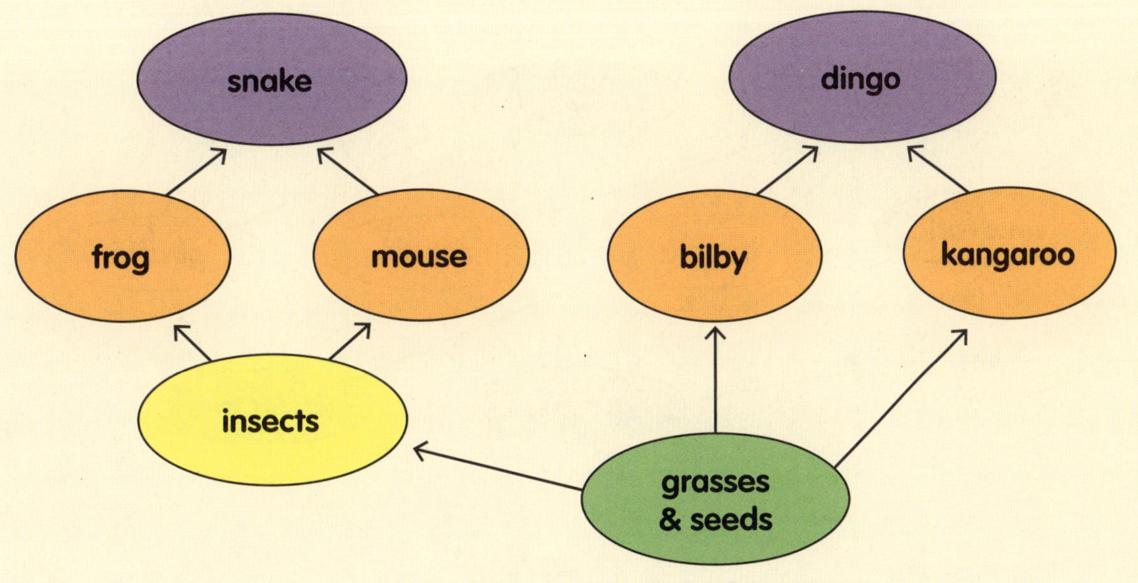

Look at the diagram of the food web and answer the questions.

1. What are the two top predators? ______________________

2. Which organisms are eaten by insects and kangaroos? ______________________

____

3. Which two animals eat insects? ______________________

Concepts:

Organisms are in competition with each other for food.

Organisms belong to many food chains that make up a food web.

Food Webs

Food Chains and Food Webs of the African Savanna

Skill:

Interpret information in graphic representations.

Write a sentence explaining the African savanna food chain diagram below. Use the term *food chain* in your sentence.

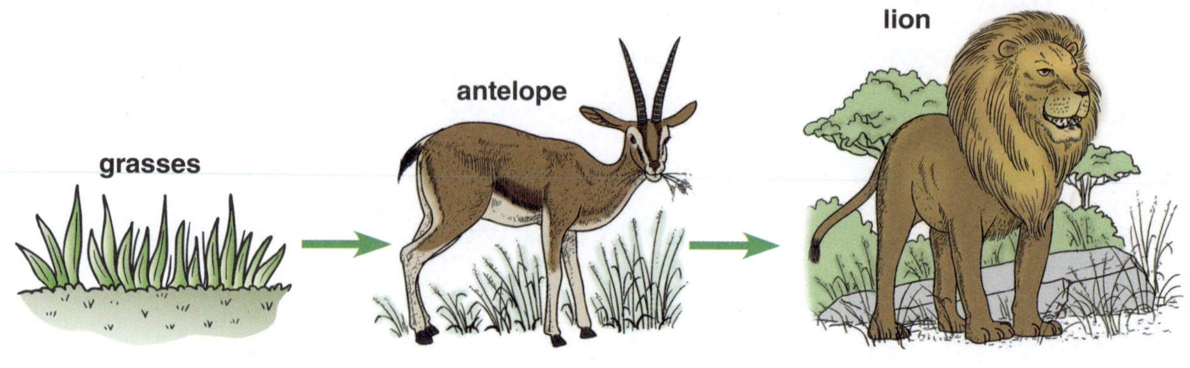

____

____

What food chains do you see in the food web below? Fill in the blanks to show three different chains.

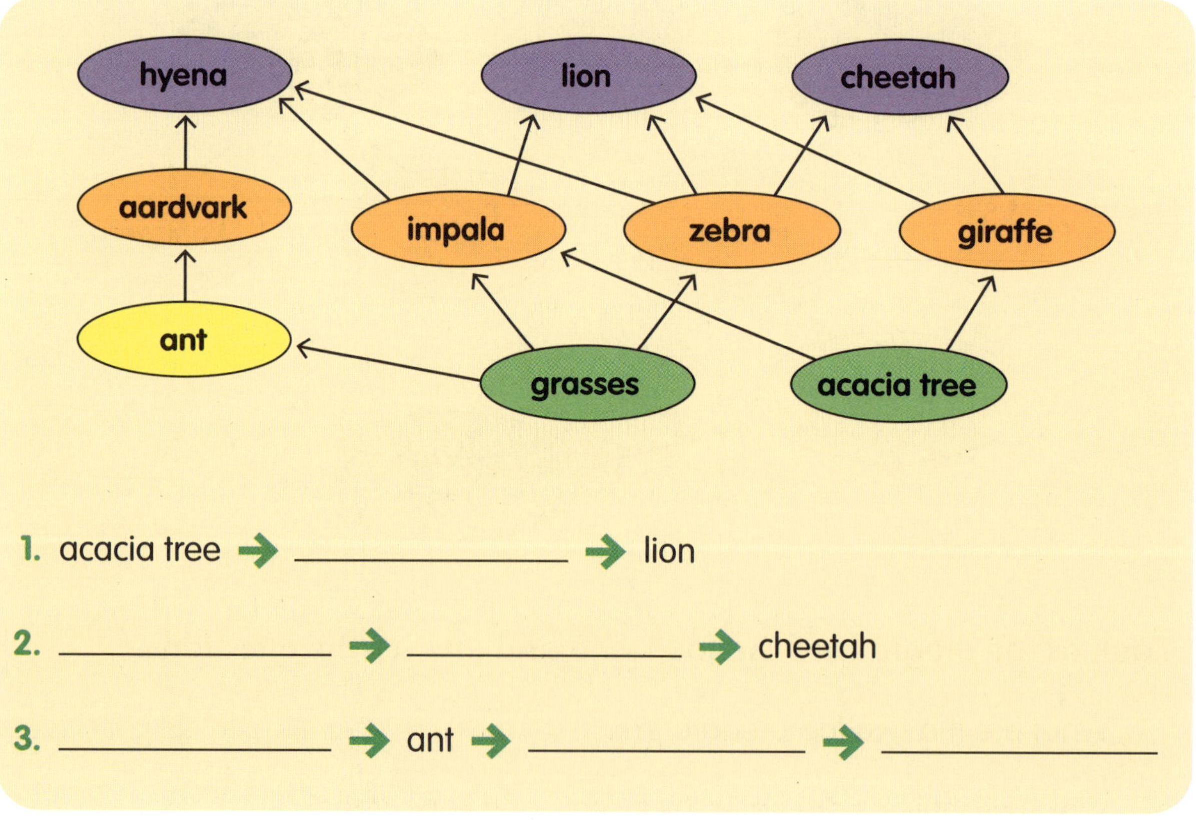

1. acacia tree → ________________ → lion
2. ________________ → ________________ → cheetah
3. ________________ → ant → ________________ → ________________

TARGETING SCIENCE YEAR 4 © PASCAL PRESS ISBN: 978125726534

Skill: Apply content vocabulary.

Select from the list of vocabulary words to complete the crossword puzzle.

carnivore	herbivore	consumer	decomposer	prey	food chain
food web	competition	omnivore	predator	ecosystem	producer

Across

3. an animal that eats only plants
7. an animal that eats both plants and other animals
8. an organism that makes its own food
9. a series of organisms in which each member feeds on the one below it
10. an organism that relies on plants and other animals for food

Down

1. an animal that hunts and kills other animals
2. an animal that only eats other animals
4. an animal hunted or caught for food by another animal
5. a struggle between two or more organisms for a limited resource
6. a network of interconnected food chains

TARGETING SCIENCE YEAR 4 © PASCAL PRESS ISBN: 978125726534

Pond Food Web

Skill:

Create graphic representations of scientific concepts.

Using these photos and descriptions, create a pond food web on the next page (page 9). Label each organism in the food web. Then draw arrows to indicate which organisms eat which other organisms. Remember to draw the arrow from the organism being eaten to the organism that is eating it.

sedge

Eaten by: maned goose, moth, ant

dragonfly

Eats: moth
Eaten by: frog

frog

Eats: moth, ant, dragonfly
Eaten by: kookaburra

kookaburra

Eats: frog

algae

Eaten by: tadpole

moth

Eats: sedge
Eaten by: lizard, frog, dragonfly

maned goose

Eats: sedge

tadpole

Eats: algae

ant

Eats: sedge
Eaten by: lizard, frog

lizard

Eats: ant, moth

Food Webs

TARGETING SCIENCE YEAR 4 © PASCAL PRESS ISBN: 978125726534

Create the Food Web

Create an Ecosystem

Skill:

Write narratives to develop real or imagined events.

Imagine that you have found a new planet where there is life. Describe one of the ecosystems on this planet. Name producers and consumers, predators and prey, and tell about how they interact with one another. Use as much detail as you can.

TARGETING SCIENCE YEAR 4 © PASCAL PRESS ISBN: 978125726534

The Soil Ecosystem

Concepts:

Soil is an ecosystem home to millions of micro-organisms.

Earthworms are decomposers.

Define It!

microorganism: an organism that can only be seen under a microscope

network: a group or system of interconnected things

recycle: to convert waste into reusable material

There's a lot more than just dirt in a garden. A closer look might reveal ants and centipedes, or perhaps a **network** of plant roots. And although you might not have seen them, the soil is full of **micro-organisms**. Thirty grams of soil can contain 100,000 algae, 1,000,000 fungi, and 100,000,000 bacteria!

You could say healthy soil is "alive", crawling with worms, insects, and microscopic life. Soil is an ecosystem that includes not only the minerals of the dirt but also all the organisms that make the soil their habitat.

In the soil ecosystem, earthworms play the role of decomposers. Earthworms break down and **recycle** matter mostly from dead plants.

Answer the questions.

1. Name the living things in the soil ecosystem mentioned in the passage.

____

____

2. What role do earthworms play in the soil ecosystem?

____

Earthworms Help the Soil

Concepts:

Earthworms' castings and burrowing help enrich the soil.

Define It!

castings: the waste that comes out of an earthworm's body

decaying: rotting

Earthworms keep the soil healthy by eating and by getting rid of their waste. Earthworms eat dead and **decaying** plant matter that falls into the soil. They also graze on microorganisms in the soil, such as bacteria and fungi. When earthworms eat, they swallow small pieces of dirt along with plant and animal matter. The worms break down this material into smaller pieces. Then they expel it in the form of **castings**, which are rich in minerals and nutrients that benefit plants and other organisms.

There are other ways earthworms help the soil. In addition to enriching the soil with their castings, they burrow, or dig down, into the soil. This allows oxygen to enter the dirt. By burrowing, the worms bring organic-rich top layers deeper down into the soil. Burrowing also helps water drain into the soil so that it may reach plant roots.

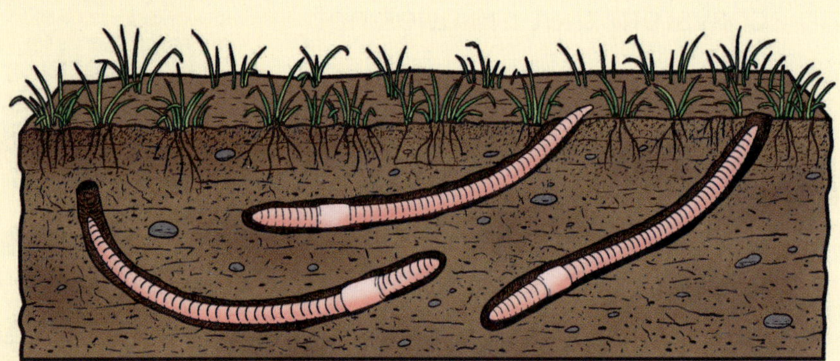

Write *true* or *false*.

1. When earthworms burrow, they remove nutrients from the soil. __________

2. Castings contain nutrients that other organisms use to survive. __________

3. Earthworms prevent water and oxygen from getting too deep in the soil. __________

TARGETING SCIENCE YEAR 4 © PASCAL PRESS ISBN: 978125726534

Concepts:

Decomposers, producers, and consumers all work together to keep an ecosystem healthy.

Define It!

consume: to eat or drink

contribution: the part played by a person or thing in bringing about a result

Remember that every ecosystem includes decomposers, producers, and consumers that work together to transfer energy. Decomposers such as earthworms are just one part of the soil ecosystem. Earthworm castings add nutrients to the soil. Nutrients in the soil allow plants to grow. Plants provide food for insects and animals that live above the soil. Animals, in turn, are **consumed** by other animals. Ultimately, the wastes and remains of plants and animals return to the soil. In this way, all the plants and animals make their **contribution** to the soil ecosystem.

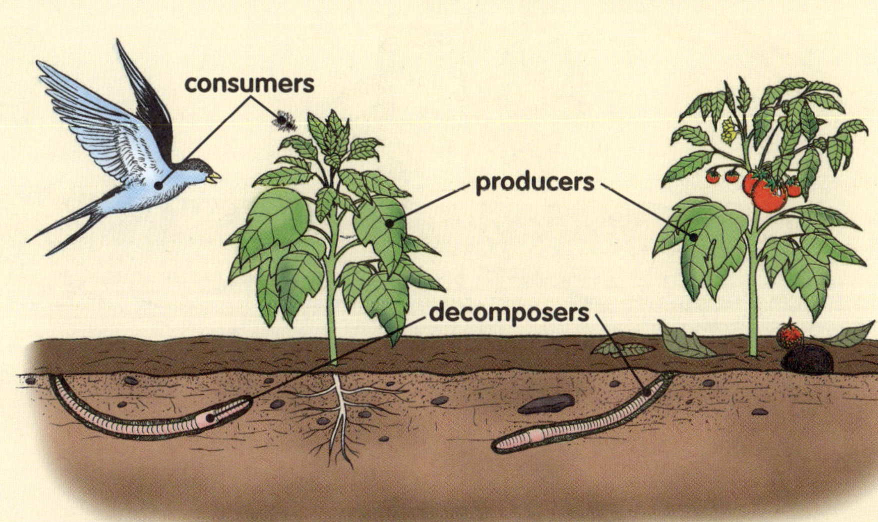

Number the events below in the correct order to show the cycle that occurs in a soil ecosystem. The first step is given.

_____ Animals and insects eat plants.

_____ Nutrients help plants grow.

__1__ Worm castings enrich the soil with nutrients.

_____ Worms decompose dead material.

_____ Animals and plants die.

A Closer Look at the Soil

Skill:

Label images that represent scientific concepts.

Producers, consumers, and decomposers all work together to keep a soil ecosystem healthy and balanced. Label the *producer*, *consumer*, and *decomposer* in this illustration. Then write a paragraph to explain how they all work together.

Decomposers

TARGETING SCIENCE YEAR 4 © PASCAL PRESS ISBN: 978125726534

Skill:
Apply content vocabulary.

Select from the list of vocabulary words to complete the crossword puzzle.

network	contribution	recycle	castings
consume	microorganism	decaying	

1 2 3 4 5

Across

1. to eat or drink
5. rotting

Down

1. the waste expelled by earthworms
2. an organism that can only be seen under a microscope
3. to convert waste into reusable material
4. a group or system of interconnected things

TARGETING SCIENCE YEAR 4 © PASCAL PRESS ISBN: 978125726534

Movers and Shakers

Skills:

Conduct experiments, record data, and analyse results.

Here's how to make a mini-ecosystem for earthworms, using a soft drink bottle and a little creativity.

What You Need

- 4 earthworms from a garden (or buy at a pet store or bait and tackle shop)
- dampened sand
- moist, light-coloured mulch
- dark, moistened potting soil
- 2 2-litre soft drink bottles with the top third of each bottle cut off (save the tops)
- shredded lettuce
- tape and aluminium foil

Directions

1. Cut off the top third of each bottle. Pour the sand, then the mulch, and then the potting soil into each bottle as shown.
2. Place the shredded lettuce on top of the soil.
3. Place the earthworms in one of the bottles.
4. Tape the top portion back onto each bottle.

5. Cover both bottles with aluminium foil and keep them in a cool place away from sunlight.
6. What you think will happen in each bottle? Why?
7. Check your bottle ecosystems every day over the next week. Record your observations on page 17. Keep the soil moist. Add fresh lettuce every 3 days.

TARGETING SCIENCE YEAR 4 © PASCAL PRESS ISBN: 978125726534

	Bottle with Earthworms	Bottle Without Earthworms
Day 1		
Day 2		
Day 3		
Day 4		
Day 5		
Day 6		
Day 7		

What Did You Discover?

1. Compare what happened to the lettuce in the two ecosystems.

____

____

2. Compare what happened to the soil layers.

____

____

3. Based on your observations, how do you think earthworms benefit the soil?

____

____

____

____

Compost It

Skills:

Conduct experiments, record data, and analyse results.

Many of the things that we throw away can be reused, recycled, or composted. Composting is when people mix yard waste such as grass clippings and leaves with leftover food to make a mixture that decomposes. This compost mixture is used to enrich garden soil.

For the next week, make a log of all the things you put in the garbage after each meal. List the foods that you think could be composted (for example, leftover food that is too old to eat, banana peels, potato skins, etc.). Then answer the questions.

1. How much food did you throw away at the end of every meal?

___

2. If you were to make a compost pile with the food you threw away, what would it be made of?

___

3. How would adding earthworms to your imaginary compost pile help it?

___

4. Did keeping the log make you think more critically about the materials you put in the garbage? Why or why not?

___

Challenge!

Research the best materials to use in order to create a nutrient-rich compost. Then use your family's food and yard waste to make your own. When the compost is ready, use it in your yard or garden, or for your potted plants!

TARGETING SCIENCE YEAR 4 © PASCAL PRESS ISBN: 978125726534

Iguana in the Ocean

Define It!

affect: to make a difference to

algae: tiny water plants

ecosystem: a community of living and non-living things that have an effect on each other

habitat: a home in nature

iguana: a type of lizard

An **ecosystem** is made up of non-living and living things in their **habitats**. When one thing changes, it **affects** the others.

The Galápagos (guh-LAH-puh-gohs) Islands are a special ecosystem because many of the animals that live there are not found anywhere else. One such animal is a lizard called the Galápagos Marine **iguana**. All lizards are land animals except for these iguanas. They swim into the ocean to eat tiny **algae** that grow there. Then they return to land to warm up in the sun. During some years, the ocean gets warmer and there is less algae to eat. When this happens, the iguanas are affected. The bones in their bodies shrink in size. Their smaller bodies warm up in the sun more quickly, so they can make more trips to feed in the water. When the ocean gets colder again, the iguanas grow larger.

Concepts:

Ecosystems include living and non-living things that interact with each other.

Non-living things can make changes in a habitat that affect the living things there.

Answer the questions.

1. What happens in the iguana's habitat that changes its food supply?

   ____

2. How do the iguanas change? ______________________________

Brown Tree Snakes

Concepts:

Humans have unintentionally changed some ecosystems by bringing in other species.

A living thing can make changes in a habitat that affect the organisms living there.

Define It!

cargo: goods carried on a ship

extinct: no longer to be found living

military: armed forces

species: a group of plants or animals that has many common traits

Without knowing it, people changed the ecosystem on the island of Guam (gwahm). Brown tree snakes slipped in more than 60 years ago. They were hiding in the wheels of **military** planes and on **cargo** boats. Now there are more than two million brown tree snakes on Guam. The snakes have caused big changes to the ecosystem. There used to be many forest birds, but the snakes ate them. Now birds are nearly all gone, and some **species** of birds are **extinct**. Because there are fewer birds to eat spiders, there are many more spiders. There are not enough birds to help spread seeds, so fewer new trees and plants grow in the forest. Scientists are working to find ways to get rid of the snakes without harming other parts of the ecosystem.

Circle *true* or *false*.

1. There are millions of snakes on Guam. **true** **false**
2. Birds are nearly gone from Guam. **true** **false**
3. Spiders are extinct species on Guam. **true** **false**

TARGETING SCIENCE YEAR 4 © PASCAL PRESS ISBN: 978125726534

Paperbark Trees

Define It!

control: to reduce the number

native: belonging to a place

survive: to continue to live

wildfire: a fire that destroys a wide area

People brought paperbark trees from Australia to Florida, and the trees changed Florida's ecosystem. People planted the trees in Florida in the 1880s, hoping the trees could help dry out swampy land. But the paperbark trees took over, blocking out light and making it impossible for Florida's **native** plants to grow. Without the native plants, many animal and insect species could not survive. Unfortunately, very few animal species can live in the paperbark tree's habitat.

Another problem is that the trees **survive** Florida's **wildfires** but Florida's native plants do not. So, there are fewer native plants and more paperbark trees. It is a very big problem to **control** the spread of these trees. Scientists know that this tree changes the ecosystem in Florida and makes native plant and animal species disappear.

Answer the questions.

1. Why are Florida's native plants important? ____________________

____

2. What problem do scientists have? ____________________

____

Concepts:

Humans have deliberately changed ecosystems by bringing in other species.

New species in an ecosystem can affect the survival of native plants and animals.

Birds on the Move

Skills:

Read and interpret data from a bar graph.

Solve single-step problems using information presented in a bar graph.

The Earth's climate is changing. As Earth grows warmer, the ecosystems in which birds find food are changing, too. For forty years, scientists have kept track of where birds live. By 2010, the habitats of birds in America had moved further north. This graph shows the distance in kilometres.

Look at the graph and answer the questions.

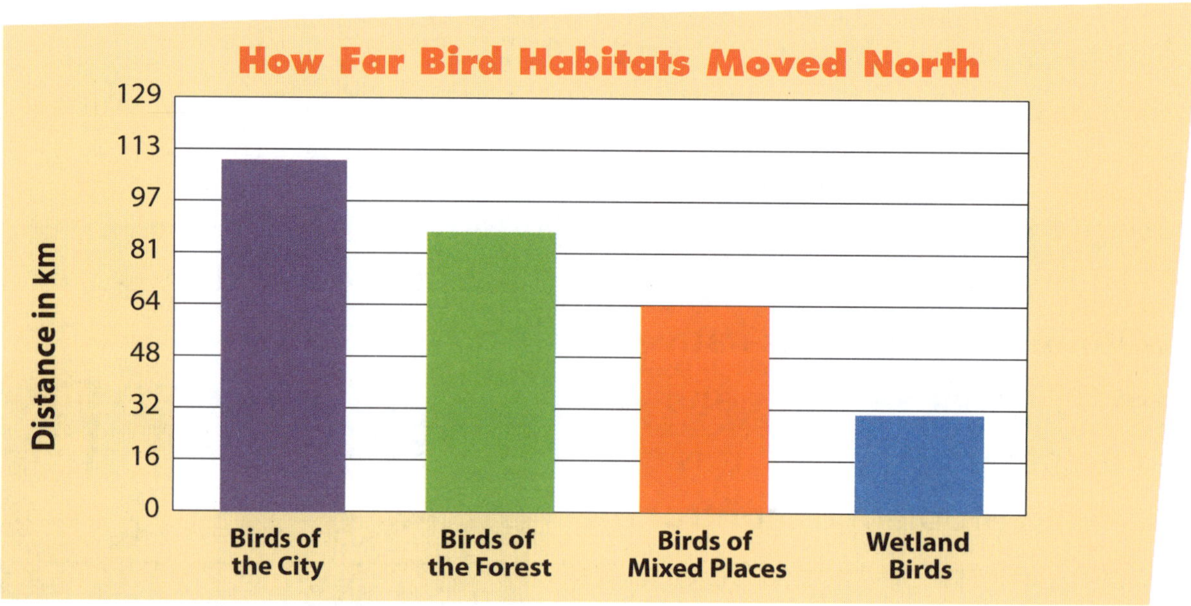

1. Which type of bird had to move the greatest distance to find food?

____

2. Which type of bird had to move about 87 km to the north?

____

3. How many kilometres did birds of mixed places have to move?

____

4. Which birds moved fewer than 32 km?

____

Ecosystems

TARGETING SCIENCE YEAR 4 © PASCAL PRESS ISBN: 978125726534

Ecosystems Crossword Puzzle

Skill: Apply content vocabulary.

Use the vocabulary words to complete the crossword puzzle.

iguana	wildfire	species	affect
native	extinct	ecosystem	algae

Across

2. tiny water plants
3. a fire that destroys a wide area
6. a group of plants or animals that has many common traits
8. no longer to be found living

Down

1. belonging to a place
4. a type of lizard
5. a community of living and non-living things
7. to make a difference to

Ecosystems

Ecosystem Visitor's Guide

Skills:

Conduct research on a science topic and write a report in the form of a national parks visitor's guide.

National parks take care of many types of ecosystems. People visit national parks to see mountains, wetlands, deserts, and forests. Make a visitor's guide for a national park you would like to visit.

Some National Parks

- Blue Mountains National Park (NSW)
- Kakadu National Park (NT)
- Daintree National Park (QLD)

What You Need

- 5 sheets of paper and 1 sheet of construction paper
- stapler
- crayons, coloured pencils, or markers
- library books or the Internet

What You Do

1. Choose a national park from the list or one of your choice. Use library books or the Internet to find out about the animal and plant species that live there.
2. Fold the sheets of paper and the construction paper in half. Staple them at the fold to make a book.
3. On each page of your visitor's guide, draw or show a picture of a plant or an animal that lives in the park.
4. Write the name of the organism and tell something about its ecosystem.

Ecosystems

TARGETING SCIENCE YEAR 4 © PASCAL PRESS ISBN: 978125726534

Galápagos Paper Mosaic

Skills:

Apply observation skills to create a paper mosaic.

Follow a sequence of directions to complete a science project.

What You Need

- construction paper (blue, black, brown or tan, green, and other colours)
- scissors
- pencil
- glue
- iguana picture, page 19

What You Do

1. Trim a half sheet of blue paper for the ocean and glue it to a sheet of tan or brown paper. This will make the land and ocean.
2. With a pencil, lightly draw a large, simple outline of an iguana. Show that this iguana lives both on land and in the sea. Use the picture on page 19 to spark your ideas.
3. Cut small squares of coloured paper. Fill in each area of the iguana by gluing on the squares. Cut smaller pieces if needed. **Hint:** It is easier and neater to put glue on the background paper rather than on the paper squares.
4. Cut paper triangles and glue them along the iguana's back.
5. Add some green algae in the ocean for the iguana to eat.
6. Add a rock or a crab to the ecosystem if you wish.

Habitat, Sweet Home

Skills:

Apply observational skills to a familiar ecosystem.

Write informative text to examine a topic and convey ideas and information clearly.

Hint

Ecosystems may have many habitats, or animal homes.

Write a list of the animals you have seen in your neighbourhood, such as birds, insects, and mammals.

______________________ ______________________

______________________ ______________________

______________________ ______________________

1. Which animal do you like best?

____

2. What is its habitat?

____

____

3. Draw a picture of the animal in its habitat.

TARGETING SCIENCE YEAR 4 © PASCAL PRESS ISBN: 978125726534

Salty or Fresh?

Almost three quarters (3/4) of the Earth's surface is covered by water. Most of this water is salty, and a very small amount is fresh (3%). This can be represented by these diagrams:

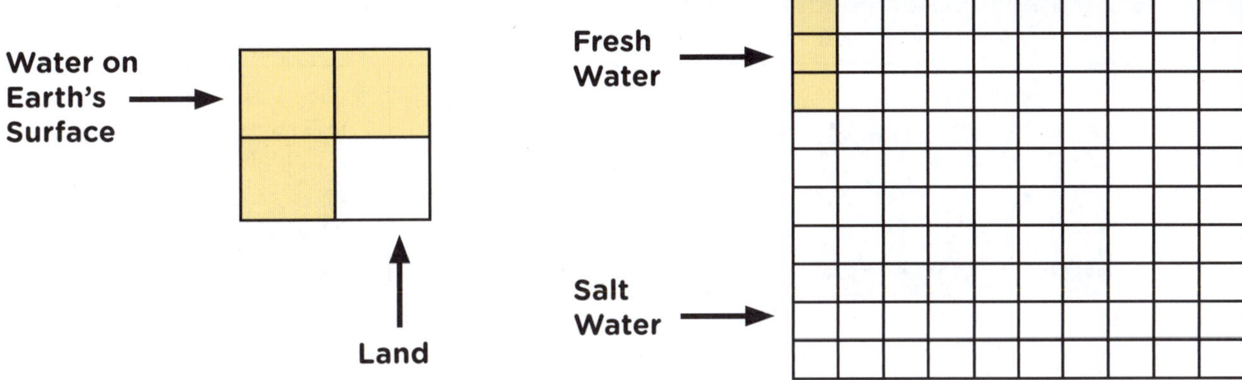

Make a list of all the things you have ever used water for. To get started, you could think about sports you've played, fun you have had, the way your food is prepared and the things you use it for, just to stay healthy.

• Drinking (F)			
• Surfing (S)			

Could you fill all the boxes? Now, look at your list, and decide which of these things you might use fresh water for, and which you would use salt water for.

1. Which type of water do you use the most of?

____

2. Where is that water really coming from?

____

3. Will it ever run out?

____

Bodies of Water

'Bodies of Water' is the title we give to the places on Earth's surface where water is found. These are just a few examples:

Puddle	Pond	Creek
 Very small pools of muddy water formed from rain.	 A small body of still, fresh water.	 A relatively small stream of flowing water.
Swamp	**Lake**	**River**
 A type of wetland area permanently filled with water.	 Runoff from local streams and rivers collecting in low lying areas.	 Larger bodies of water that flow downhill from creeks or streams.
Harbour	**Sea**	**Ocean**
 Any type of body of water sheltered from the ocean waves.	 Part of the ocean that is at least partially bordered by land.	 The largest bodies of water on Earth.

1. Which would you be most likely to do these activities in?

   Somewhere to put some pet goldfish: ____________

   A safe place to moor a very large ship: ____________

   A place crocodiles and snakes might like to hide: ____________

   Something you can jump in after rain: ____________

   A safer place to go canoeing: ____________

   A place where you might find giant cargo ships: ____________

TARGETING SCIENCE YEAR 4 © PASCAL PRESS ISBN: 978125726534

Changes to Bodies of Water

The amount of heat or rainfall we get can change the surrounding water bodies in many ways. Look at each of these water bodies and suggest how each change would effect them.

Example:

Puddle:	After very a very hot day (drawing):	Explain what would happen?
		The water would dry up from evaporation.

Try these:

Sea:	Heavy rain and strong wind:	Explain:

Swamp:	**Very heavy rain:**	**Explain:**

River:	**No rain for a really long time:**	**Explain:**

Water

Tap Water

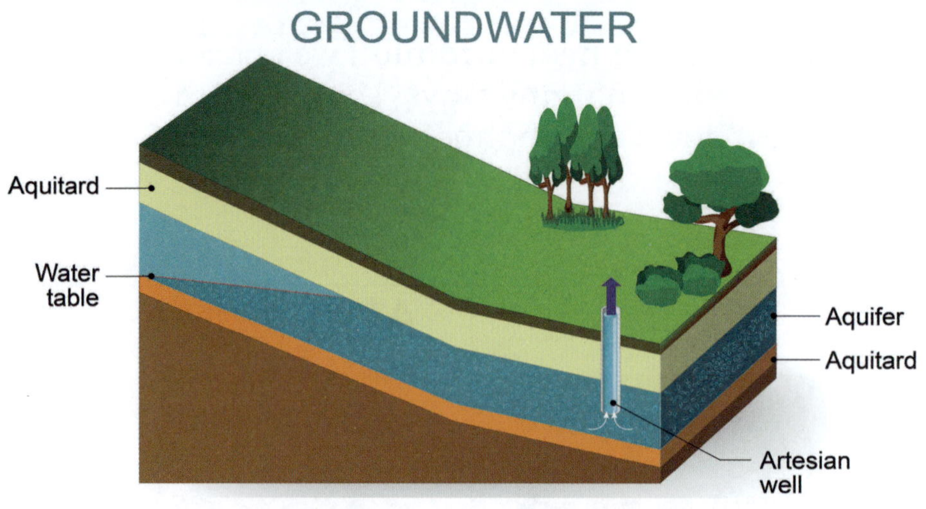

In Australia, we collect our water from three main sources:

Groundwater: Water that comes from the rain and soaks down into the layers under the ground.

Surface Water: Rain drained into rivers or creeks or collected in dams and water tanks.

The Ocean: Water that is taken from the ocean and has the salts and minerals removed.

Before water can be used for drinking or washing it must be 'cleaned' at a water treatment plant through several treatment processes. Mixing alum (aluminium sulphate) with water and allowing it to settle removes mud, dirt and other particles.

Sand and gravel-filled filters remove tiny particles and chlorine is added to kill bacteria. The treated water is pumped to reservoirs for storage.

Reservoirs are usually on high ground so that water can flow into underground pipes or water mains. The water in the mains flows into the house when you turn on the tap.

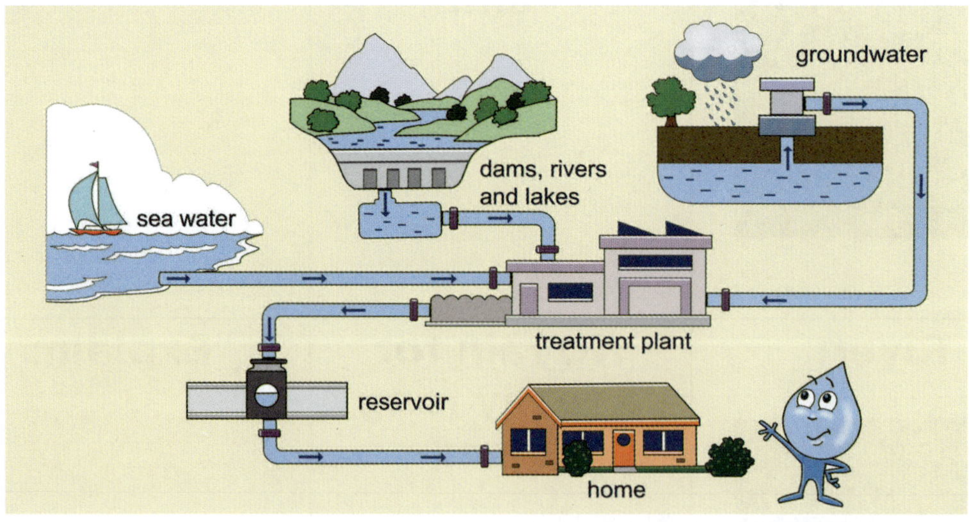

1. What do you think happens to water when it goes down the drain?

____

____

This chart shows some suggestions for saving water. Can you think why each of these actions could help us to use less water? You may need to research how water is used in some of these things. Are there any of these ideas you could do to use less water? Can you think of others that aren't included in the chart?

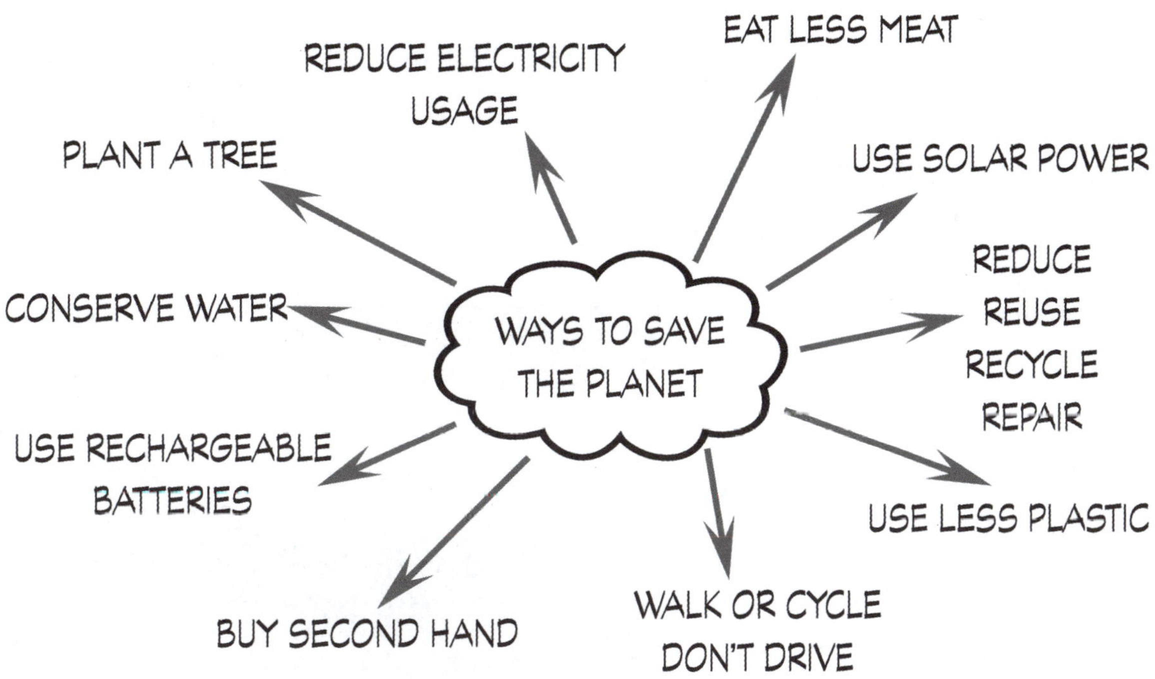

Things I could do to save water:

Turn off the tap when I clean my teeth.		

Valuing Water

For First Nations peoples living in places where fresh water was scarce, traditional knowledge of the local places where water could be found was passed down from generation to generation through instructions, stories and paintings, including, at times, stylised mapping.

This knowledge was really important for their survival. Many Torres Strait Islander communities have, for thousands of years, successfully managed difficult fresh-water supply issues, carefully collecting rainwater in a variety of ways.

In desert areas, where rainfall is very low, First Nations peoples draw on their knowledge of a variety of water sources to meet their needs. These sources include riverine waterholes, soakage-wells in areas where water seeps down into the ground, flooded rock holes (known as gnammas, a Nyungar word from south-west Western Australia), rainwater accumulated in tree hollows and even water from the body of the water-holding frog!

On Damut Island in the Torres Strait, people used sticks and blocks of wood to stop the water in waterholes from drying out. The Baiyungu People of the north west Australian coast protected their water supplies after rainfall by covering the water that collected in rock cavities using flat pieces of limestone.

In the Crystal Brook region of South Australia, the Nukunu People prevented evaporation of water from a deep spring by covering it with bushes and branches from trees.

To prevent water drying out in rock-holes, the Yankuntjatjara Peoples placed sand in the holes. This covered the water and protected it. To get to the water, a hole was dug in the sand, and the fresh water filled the hole.

1. If you were trapped in the desert, where are some places you could perhaps look for water, using some of the ideas passed down for thousands of years?

____

____

____

____

____

Precipitation

Concepts:

Water in the atmosphere forms clouds.

Water falls back to Earth in the form of precipitation.

The atmosphere holds water in the form of **water vapour**. Water vapour is made when water **evaporates**, or changes from a liquid to a gas. The vapour disappears into the air. If water vapour is lifted high enough into the air, it cools. Cooling causes the water vapour to **condense**, or form tiny drops of water. This is how clouds are formed. The tiny drops grow and change inside the clouds until they become so heavy that they fall to the ground as **precipitation**. Rain, snow, sleet, and hail are all types of precipitation. Scientists use a **rain gauge** to measure the rain that falls in a place.

Define It!

condense: to change from a gas to a liquid form

evaporate: to change into a gas

precipitation: water that falls as rain, snow, sleet, or hail

rain gauge: a tool that measures rainfall

water vapour: the gas that clouds are made of

rain snow sleet hail

Answer the questions.

1. How does water get into the air? ______________________

2. What happens to water vapour when it cools?

____

Weather

 TARGETING SCIENCE YEAR 4 © PASCAL PRESS ISBN: 978125726534

A Cloud Record

Skills:

Make observations to gather data and present it in a table.

Scientists group clouds into three main types. You can tell each type of cloud by its shape and by how high it is in the atmosphere.

Stratus (STRAY-tuhs)
Stratus clouds are the lowest clouds. They look like a flat grey sheet in the sky. Light rain or snow may fall from stratus clouds. Stratus clouds close to the ground are called *fog.*

Cumulus (KYOO-myuh-luhs)
Cumulus clouds are puffy white clouds, higher than stratus clouds. They mostly bring fair weather. However, they may gather and form huge storm clouds. Those bring thunder, lightning, rain, and hail.

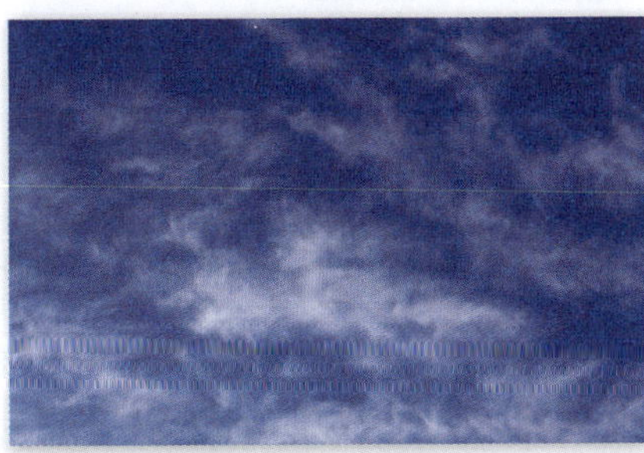

Cirrus (SEAR-uhs)
Cirrus clouds are thin clouds that look feathery. They are the highest clouds in the sky. Cirrus clouds are a sign that the weather may change.

My Cloud Record

Day	Type of Cloud	Weather Today	I Predict Tomorrow Will Be...

Weather

Cloud in a Glass

Skill:

Conduct experiments and answer questions

Make a cumulus cloud in a glass and then make it rain.

What You Need

- shaving cream
- clear glass
- water
- blue food colouring

What You Do

1. Fill the glass halfway with water.
2. Pile a mound of shaving cream on top of the water in the glass. Make it look like a fluffy cumulus cloud.
3. Drip some blue food colouring onto the cumulus cloud. Watch as the cloud absorbs the water. Soon, the cloud will get too heavy and blue drops will fall from the cloud.
4. How is this similar to how rain really happens?

____

____

Concept: Glaciers are large sheets of ice that are found in places that are cold year-round.

Define It!

glacier: a slow-moving mass of ice

pressure: the weight or force produced when something presses against something else

Have you ever seen a snow-topped mountain? If so, you may have been looking at a **glacier**. Glaciers are large sheets of ice that form in places where more snow falls than melts. As layers of snow build upon one another, the weight from the top layers presses down on the layers underneath. This **pressure** turns the snow to ice, like when you squeeze fluffy snow into a hard snowball.

Because glaciers form slowly, most are found in places that are cold year-round. These places include Greenland, Antarctica, Alaska, and the tops of mountains.

This glacier sits on top of a mountain in Chile.

Complete the sentences.

1. Glaciers can only be found in places that are ______________ throughout the entire year.

2. Glaciers form in places where more ______________ falls than melts.

Moving Ice

Concepts:

Glaciers are moving objects that can advance and retreat.

Movement of glaciers has helped to shape the land.

Define It!

advance: to move forward

massive: very large

retreat: to move backward

Glaciers might appear to stay in one place, but they are actually "rivers" of ice that flow downhill. Glaciers can **advance** and **retreat** great distances, depending on the amount of snow that has fallen or ice that has melted. When a glacier advances, it flows farther downhill or spreads out. When a glacier retreats, it moves backward. This is because the ice is melting faster than the glacier is growing.

Glaciers are the largest moving objects on Earth, scraping rocks and soil from their paths like giant bulldozers. Movement of these **massive** sheets of ice can reshape the land over thousands of years.

Look at the diagram of a glacier. The lines show how far the ice retreated between the years 1850 and 2000.

During which period of time did the glacier retreat the most? ____________________

Glaciers

TARGETING SCIENCE YEAR 4 © PASCAL PRESS ISBN: 978125726534

Glaciers Shape the Land

One way that glaciers shape the surface of our planet is by erosion. The ice carries broken rocks and soil over long distances and deposits the **debris** far from its original location. One of the best examples of this kind of erosion is California's Yosemite Valley. Huge glaciers carved a giant U-shaped valley in the rock and left behind **ridges** of dirt and gravel called **moraines**.

In other places, erosion by glaciers created lakes. The Great Lakes in the USA were formed from **basins** scooped out by moving glaciers. When the ice melted, these basins filled with water.

Define It!

basin: a large hole in the ground that can contain water

debris: small pieces of rock

moraine: a ridge of loose rock and soil created by a glacier and left behind when the glacier melts

ridge: a long and narrow raised area

Yosemite Valley

Name two famous places created by glaciers.

1. ______________________ **2.** ______________________

Concepts:

Glaciers shape the land by erosion.

Glaciers can leave behind moraines and basins.

Moraines and Basins

Skill:

Label images that represent scientific concepts.

Look at the diagram. Label the *glacier, moraines,* and *basin*.

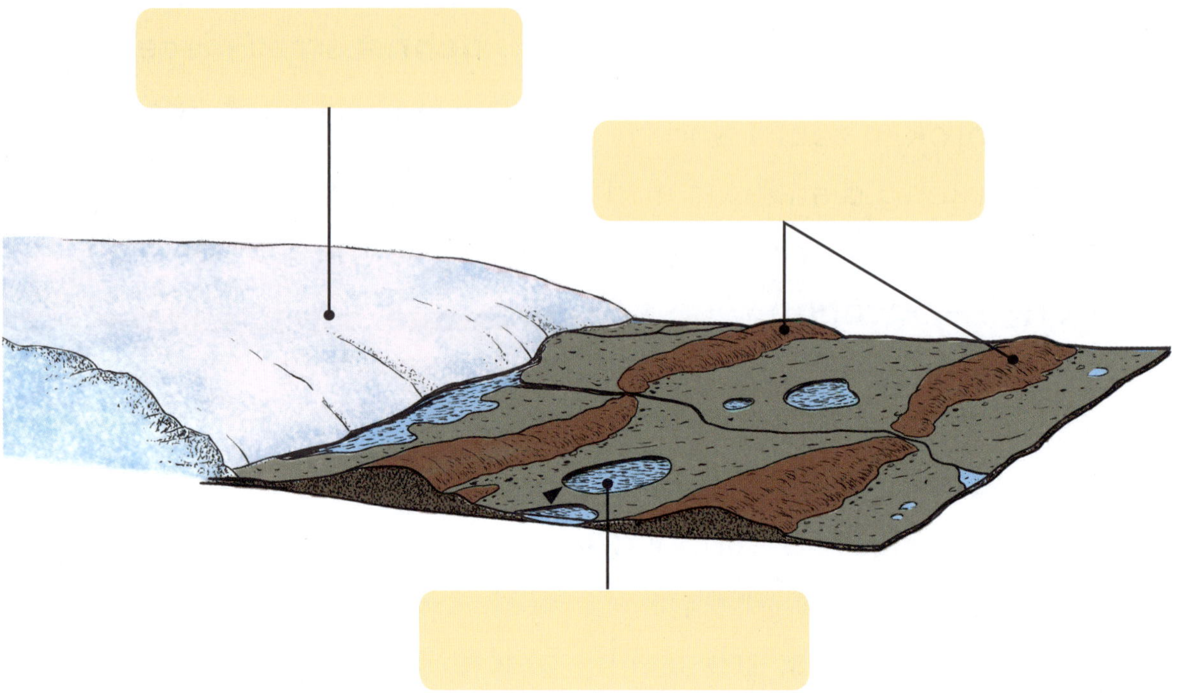

Write *true* or *false*.

1. Moraines create ridges of loose rock and soil. ____________
2. Moving glaciers can scoop out basins in the land. ____________
3. When glaciers melt, they can leave behind moraines. ____________
4. Both basins and moraines are created by erosion from glaciers. ____________
5. The Great Lakes were formed from moraines created by glaciers. ____________

TARGETING SCIENCE YEAR 4 © PASCAL PRESS ISBN: 978125726534

Skill: Apply content vocabulary.

Write each answer.

1. Is something that is **massive** very large **or** very small?

____

2. When a **glacier** moves backward, is it **advancing or retreating**?

____

3. Were the Great Lakes formed from **basins or moraines**?

____

4. Does **pressure** from the weight of the snow **or** erosion of the land **help to form glaciers**? _____________________

5. Is **debris** deposited by mountains **or** by glaciers?

____

6. When a **glacier** flows farther downhill, is it **retreating or advancing**? _____________________

7. Are **moraines** large holes in the ground **or ridges** of dirt and gravel? _____________________

TARGETING SCIENCE YEAR 4 © PASCAL PRESS ISBN: 978125726534

Grinding Glaciers

Skills:

Conduct experiments and draw conclusions about the results.

In this activity, you will look at the effect that the movement of glaciers has on Earth's surface. You will use "sandy" ice cubes as a model of a glacier that has pieces of rock in its ice.

What You Need

- ice cube tray
- water
- several handfuls of clean sand
- aluminium foil
- plastic tub
- paper to cover the table
- paper towels for cleanup

Directions

1. Make "sandy" ice cubes by sprinkling sand into an ice cube tray filled with water and freezing it overnight.
2. Smooth out a sheet of foil on the top of a table or desk.
3. Rub an ice cube across the sheet of foil. Record what you observe in the chart on page 43.

4. Stack all the ice cubes inside one corner of the plastic tub and allow them to melt. Record what you observe.

  ISBN: 978125726534

	Observations
Ice cube on foil	
Ice cubes in tub	

What Did You Discover?

1. What happened when you rubbed the ice cube across the foil?

____

2. What was left in the plastic tub after the ice cubes melted? What would this be called when a real glacier melts?

____

3. What did the experiment show you about the ways that glaciers change Earth's surface?

____

____

Melting Ice

Skill:

Write an opinion piece supporting a point of view with reasons.

Today we live in a very warm period, and glaciers are on the move—backward! Most glaciers are melting faster than they are growing, and this has scientists worried about the future of our planet.

This sign reads "The glacier was here in 1908."

Pretend that you are a scientist studying the impact of the changing glaciers on the world today. What are five questions you would ask?

Example

What might happen if all the glaciers in Antarctica were to completely melt?

1. ______________________________

2. ______________________________

3. ______________________________

4. ______________________________

5. ______________________________

TARGETING SCIENCE YEAR 4 © PASCAL PRESS ISBN: 978125726534

The Hydrosphere and the Water Cycle

https://clickv.ie/w/7xgx

Use this QR code to access a video on this topic.

Water can be found on Earth in many places—underground, in oceans and seas, and in the atmosphere. All the water that exists on and surrounding Earth is called the **hydrosphere**. The water in the hydrosphere is always moving. Water travels from ocean to air to land and back in a continuous process called the **water cycle**.

The sun causes water to move between the ocean, air, and land. The sun's energy heats liquid water and causes it to **evaporate**, or change into a gas called water vapour. Water vapour then enters the atmosphere. Most of the evaporation on Earth happens in oceans that are close to the equator. That's because the sun's heat is greatest near the equator. The wind blows the humid air from the equator long distances—it can end up anywhere in the world!

Define It!

evaporate: to change from a liquid into a gas

humid: containing a high amount of water vapour

hydrosphere: all the water on and surrounding Earth's surface

water cycle: the continuous movement of water between Earth's oceans, atmosphere, and land

equator

Concepts:

All the water in and around Earth is called the hydrosphere.

Water travels from ocean to air to land in a process called the water cycle.

Complete the sentences.

1. When liquid water ______________________ , it enters the atmosphere as a gas.

2. The movement of water between the ocean, air, and land is called the ______________________.

3. The ______________________ includes all the water on and surrounding the planet.

Condensation and Precipitation

Concepts:

Condensation is what happens when water changes from gas to liquid.

Precipitation is what happens when water becomes too heavy for the air and falls to the ground.

Define It!

condensation: the change from a vapour or gas into a liquid

precipitation: water droplets that fall to the ground as rain, hail, sleet, or snow

When water vapour moves to cooler regions—either away from the tropics or higher up into the atmosphere—it cools. As water vapour cools, it gives up its heat energy and changes back into a liquid. This process is called **condensation**. In the atmosphere, condensation takes the form of tiny droplets of water. We see condensation as clouds in the sky or fog near the ground.

When water droplets get so big that air currents can no longer hold them, they fall to Earth's surface as rain. When the air is cold enough, the water droplets freeze, and snowflakes form. Rain, snow, hail, and sleet are all forms of **precipitation**.

Write whether the phrase describes *condensation* or *precipitation*.

1. frost on the window ____________________
2. morning dew on the grass ____________________
3. snowflakes falling to the ground ____________________
4. fog forming in the valley ____________________

The Water Cycle

TARGETING SCIENCE YEAR 4 © PASCAL PRESS ISBN: 978125726534

A Continuous Cycle

Define It!

compose: to form; to make up the parts of something

Once water falls to Earth's surface as precipitation, a couple of things can happen. Some of the precipitation soaks into the ground. Some water may collect in streams, rivers, lakes, and seas. If precipitation falls as snow, it may remain frozen on the ground until temperatures change and the snow or ice melts. Ultimately, most of the water that falls as precipitation ends up in the ocean. From there the water is evaporated, and the water cycle begins again.

The processes of evaporation, condensation, and precipitation have recycled water on Earth for billions of years. In fact, every living thing is **composed** of water and is part of the water cycle, too!

Label the stages of the water cycle in the diagram below, using the words *evaporation, condensation,* and *precipitation.*

When seawater evaporates, the salt is left behind. You might think this would make the water saltier. But it doesn't. Why do you think that is?

____

____

A Continuous Cycle

Skill:

Apply content vocabulary.

Use the vocabulary words to complete the sentences. Then unscramble the shaded letters to decode the secret message.

hydrosphere   water cycle   condensation   composed
precipitation   evaporates   humid

1. The ___ ___ ___ ___ ___ ___ ___ ___ ___ ___ ___, which contains all the water on and surrounding the planet, is one of Earth's major systems.

2. The air often feels warm and full of moisture on a ___ ___ ___ ___ ___ day.

3. Sleet and hail are both forms of

___ ___ ___ ___ ___ ___ ___ ___ ___ ___ ___ ___ ___.

4. Your mirror fogging up in the bathroom after a hot shower is an example

of ___ ___ ___ ___ ___ ___ ___ ___ ___ ___ ___ ___.

5. Every living thing is ___ ___ ___ ___ ___ ___ ___ ___ of water, including human beings!

6. The ___ ___ ___ ___ ___ ___ ___ ___ ___ ___ is a continuous process that moves water from the ocean to the air to the land and then back to the ocean.

7. When you boil water, the liquid ___ ___ ___ ___ ___ ___ ___ ___ ___ ___ into a gas.

Another of Earth's major systems is the

b ___ ___ ___ ___ ___ ___ ___ ___, which is made up of all living things, including people.

The Water Cycle

TARGETING SCIENCE YEAR 4 © PASCAL PRESS ISBN: 978125726534

Complete the Diagram

Label the three processes taking place in this diagram. Write a title for the diagram. Then write a paragraph explaining what is happening to the water in the diagram.

____

Title

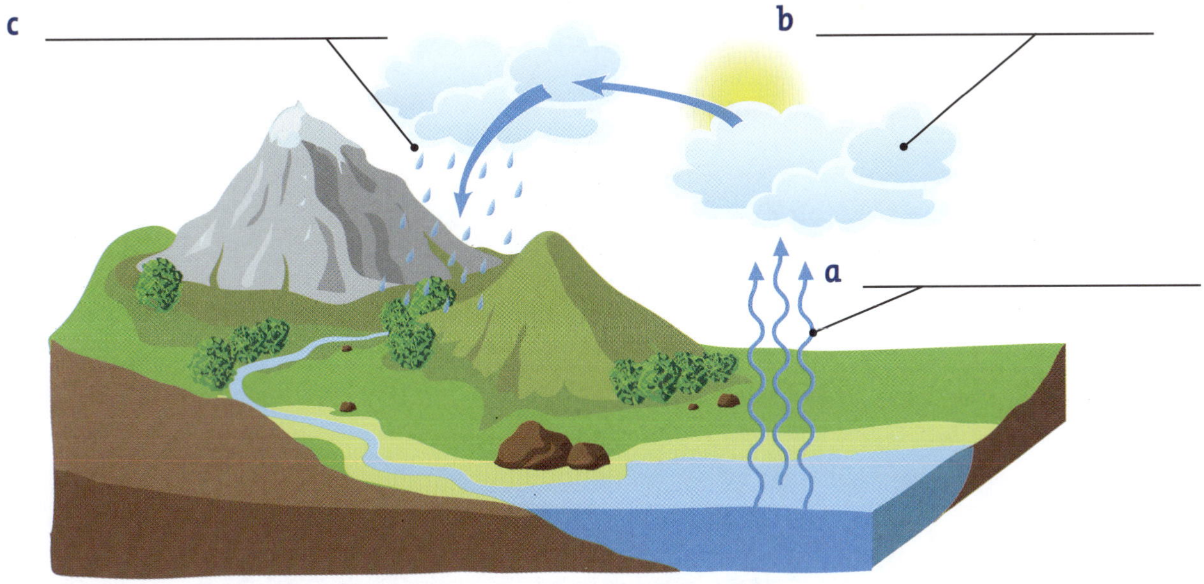

c ______________________ b ______________________

a ______________________

a ____

____

____

____

b ____

____

____

____

c ____

____

____

____

Skill:

Label images that represent scientific concepts.

The Water Cycle

Water Cycle at Work

Skills:

Conduct experiments and analyse results.

See the water cycle in action in this simple experiment using water, soil, and the power of the sun.

What You Need

- large plastic container
- small paper cup
- small rock
- plastic wrap
- wide tape
- 4 cups (544 g) of soil or sand
- ½ cup (120 mL) of water

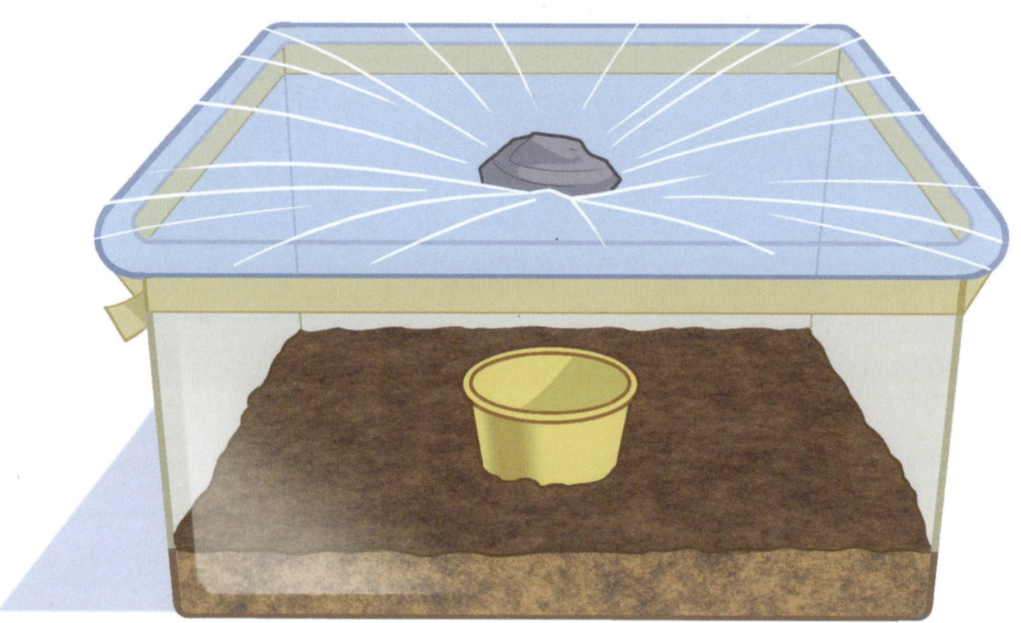

Directions

1. Pour the soil into the plastic container and spread it around evenly. Then place the paper cup in the middle of the container. Bury the cup partially in the soil.
2. Pour the water evenly over the soil, but not in the cup.
3. Cover the container with plastic wrap and seal it around the edges with tape. Place the rock on top of the wrap, centreed directly above the cup.
4. Place the container in a sunny location. Check on it the following day.

TARGETING SCIENCE YEAR 4 © PASCAL PRESS ISBN: 978125726534

What Did You Discover?

1. What happened to the water?

2. How much water did you end up with in the cup? How did it compare to the amount of water you started with?

3. Why do you think you placed the rock directly above the cup?

4. Do you think this experiment would work on a cloudy day? Explain your answer.

The Water Cycle A Water Molecule's Journey

Skill:

Write narratives to develop real or imagined experiences or events.

Pretend that you are a water molecule in the ocean. Use what you've learned about the water cycle to write a story about your journey from ocean to air to land and back again.

TARGETING SCIENCE YEAR 4 © PASCAL PRESS ISBN: 978125726534

Define It!

dissolve: to mix thoroughly with a liquid and become a solution

glacier: a huge mass of ice

Concepts:

Earth has far more salt water than fresh water.

Fresh water can be found in lakes, rivers, glaciers, underground, and in the atmosphere.

Earth is often called the Blue Planet. That's because almost three-quarters of Earth's surface is covered by water. Most of that water is salt water found in the world's oceans. Salt water contains **dissolved** minerals. You cannot drink salt water. Less than 3% of all the water on Earth is fresh water, the kind we drink.

Although you might think that most of the fresh water on Earth is found in lakes and rivers, in fact, only a small amount can be found in these places. Most of the fresh water is frozen in polar ice caps and glaciers. The rest is in the atmosphere as gas or clouds, or is located underground.

Write *true* or *false*.

1. A small amount of fresh water is frozen in polar ice caps. __________
2. Earth's oceans make up most of the water on the planet. __________
3. Three percent of Earth's water is salt water. __________
4. Some fresh water is stored in the atmosphere. __________

Fresh Water

Concepts:

Most of Earth's fresh water is in glaciers and ice caps.

The second-largest store of water on Earth is in the ground.

Define It!

groundwater: fresh water that exists underground

irrigation: supplying dry land with water artificially

transport: to move from one place to another

Remember that less than 3% of all the water on Earth is fresh water, and only a tiny fraction of that water is available for us to use. Most of Earth's fresh water is ice in the polar ice caps and in glaciers, which are far from where most people live. Unlike liquid water that can flow through pipes, frozen water is hard to **transport**. Because of this, people use fresh water from rivers and lakes near where they live.

The second-greatest source of fresh water on Earth is in the ground. This type of water is called **groundwater**. There is more fresh water available as groundwater than in all the world's rivers and lakes combined. Places that don't have fresh water from rivers and lakes depend on groundwater for drinking, **irrigation**, and other uses.

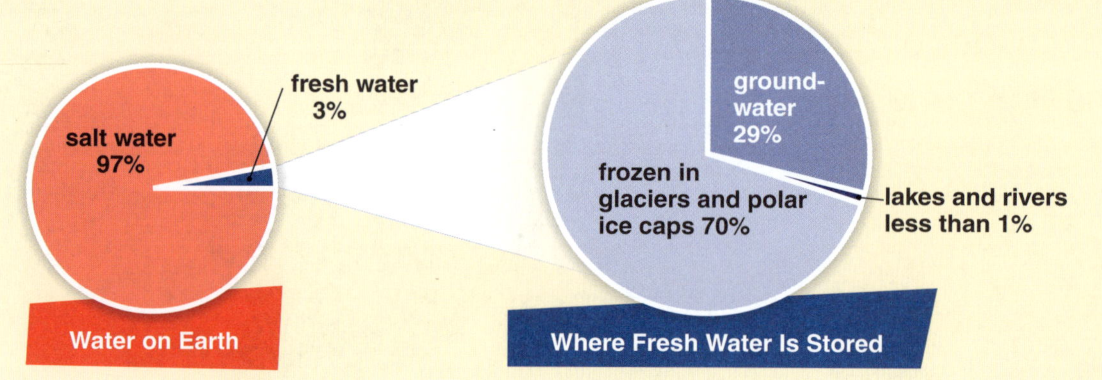

Look at the diagram and answer the questions.

1. What is Earth's second-greatest source of fresh water? ____________

2. About what percent of Earth's fresh water is found in groundwater, rivers, and lakes? ____________

3. Why is most of Earth's fresh water difficult to use?

____

TARGETING SCIENCE YEAR 4 © PASCAL PRESS ISBN: 978125726534

Threats to Groundwater

Groundwater is stored in a rock or sand layer called an **aquifer**. When it rains, water soaks into the aquifer until it reaches a certain level. This underground level is called the **water table**.

The fact that groundwater is stored underground does not protect it from **pollution**. Chemicals used in farming and waste materials from factories can seep into the ground or get carried into an aquifer by rainwater. Overuse is another threat to groundwater. When communities use too much groundwater, the water table can drop below the reach of wells. This drop can also cause seawater to flow into an aquifer, resulting in groundwater that is too salty to use.

Earth has a limited supply of usable water. There are now more than 7 billion people in the world. We each use about 45 L of water per day. Some scientists think we cannot **sustain** this rate of water usage. One of the best ways to address the problem of shrinking water supplies is to practice **conservation**. People can conserve water by using less in their everyday lives.

Define It!

aquifer: an underground layer of rock or soil that contains water

conservation: the careful use of something to preserve it

pollution: substance that makes land, water, or air dirty or unsafe

sustain: to exist; to continue

water table: the level below which the ground is full of water

Concepts:

Groundwater is threatened by pollution and overuse.

We can protect our fresh water by practising conservation.

Label the *well, aquifer,* and *water table*.

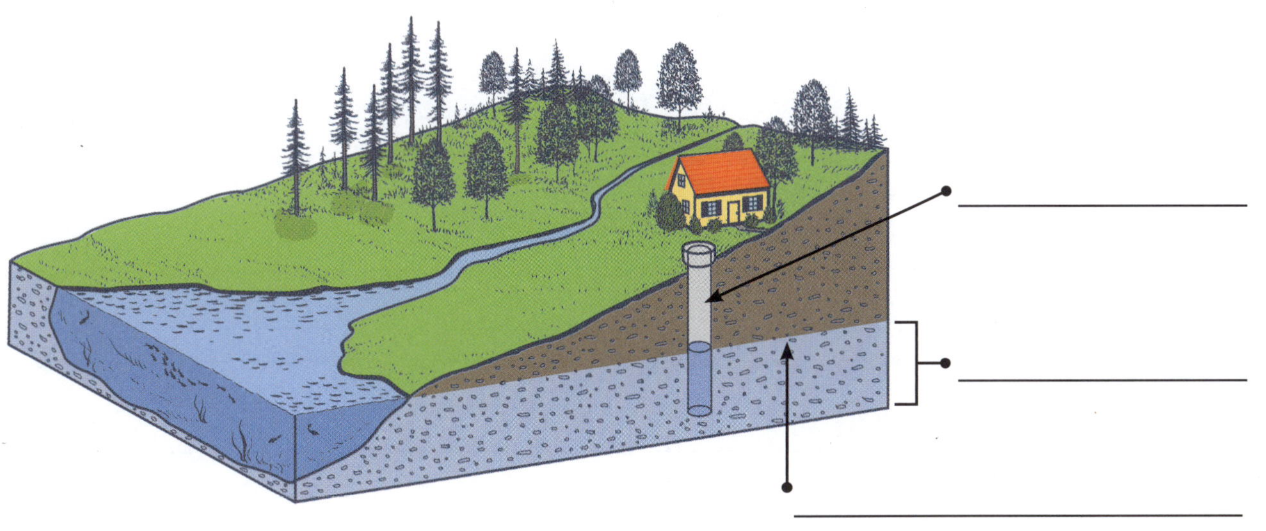

A Closer Look at Groundwater

Skill:

Interpret information from graphic images.

This diagram shows several ways in which groundwater is being threatened. Look at the labels to understand what is happening. (*Agricultural* refers to farming. *Industrial* refers to factories. *Saltwater intrusion* refers to seawater entering the aquifer.)

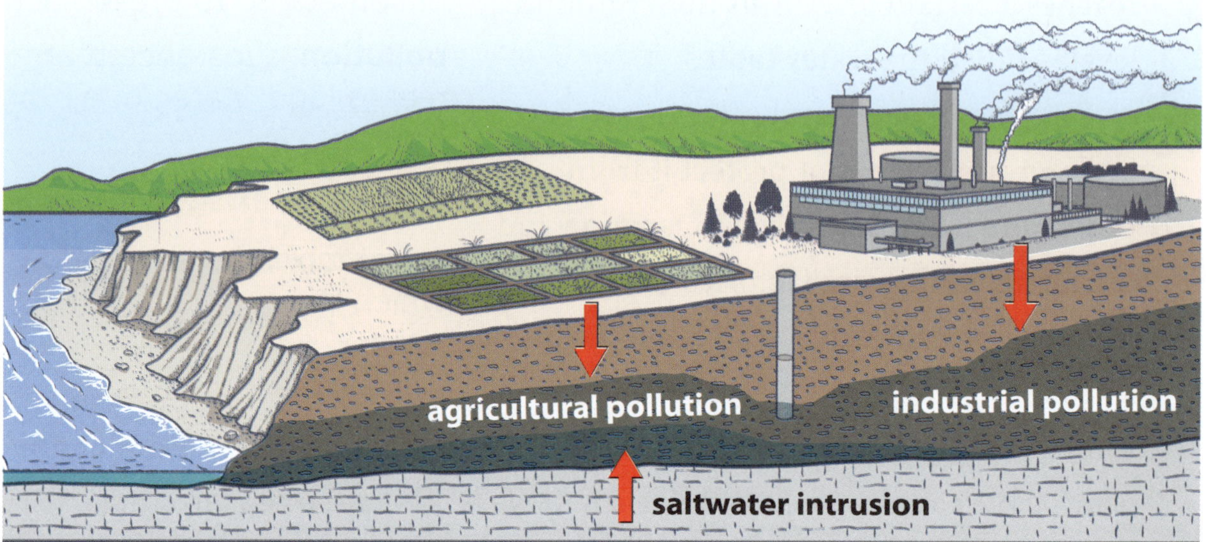

Describe the ways in which groundwater is being threatened in this diagram.

____

____

____

____

____

____

____

____

____

Skill: Apply content vocabulary.

Select from the list of vocabulary words to complete the crossword puzzle.

aquifer	pollution	sustain	dissolve	glacier
groundwater	transport	conservation	irrigation	water table

Across

2. to supply dry land with water artificially
5. to exist; to continue
7. underground level below which the ground is full of water
8. fresh water that exists underground

Down

1. substance that makes land, water, or air dirty or unsafe
3. the careful use of something to preserve it or keep it from running out
4. the process of mixing and becoming a solution
6. an underground layer of rock or soil that contains water

Build Your Own Aquifer

Skills:

Conduct experiments, record data, and analyse results.

Remember that groundwater is stored in a layer of rock or soil called the *aquifer*. People obtain groundwater by drilling wells into aquifers and pumping out the water. In this experiment, you will create your own aquifer and see the impacts of pollution on groundwater.

What You Need

- 7 x 8.25 cm clear plastic cup
- piece of modelling clay or floral clay
- white play sand
- ½ cup (115 g) aquarium gravel or small pebbles
- red food colouring
- bucket of clean water
- small cup

clay

sand

Directions

1. Put ½ cm of white sand in the bottom of the cup. Pour water into the sand, wetting it completely. Record your observations on page 59.
2. Cover half of the sand with clay. Flatten the clay like a pancake and press part of it on one side of the cup as shown above.
3. Pour a small amount of water onto the clay. Record your observations.
4. Use the aquarium rocks to create the next layer. Place the rocks over the sand and the clay. Make a slope with the rocks, piling them high on one side as shown. This will create a small hill and a valley.
5. Pour water over the rocks until it reaches almost the top of the rocks. Record your observations.
6. Put a few drops of food colouring on top of the rock hill as close to the wall of the cup as possible. Record your observations.

TARGETING SCIENCE YEAR 4 © PASCAL PRESS ISBN: 978125726534

Build Your Own Aquifer

Describe what happens to the water and/or food colouring at each step.

	Observations
Water on white sand	
Water on clay	
Water on rocks	
Food colouring in cup	

What Did You Discover?

1. In which layer did the water not soak into the aquifer? Why do you think it did not?

____

2. Water that stays aboveground is called surface water. Which layer of this experiment represents surface water?

____

3. Where is the water table in this aquifer?

____

4. How do you think pollution that happens aboveground can impact water below it?

____

Water in the Home

Skill:

Interpret information from graphic images.

This pie chart shows the daily percentage of water that is typically used in homes. Use the chart to answer the questions.

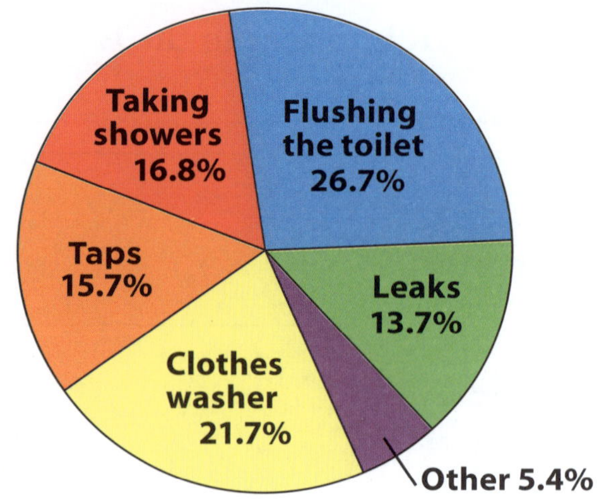

1. In which room of the house do people use the most water?

   ____

2. According to the chart, "Other" activities account for more than 5% of people's home water use. What might these be? Name two other ways you use water besides the ways listed in the chart.

   ____

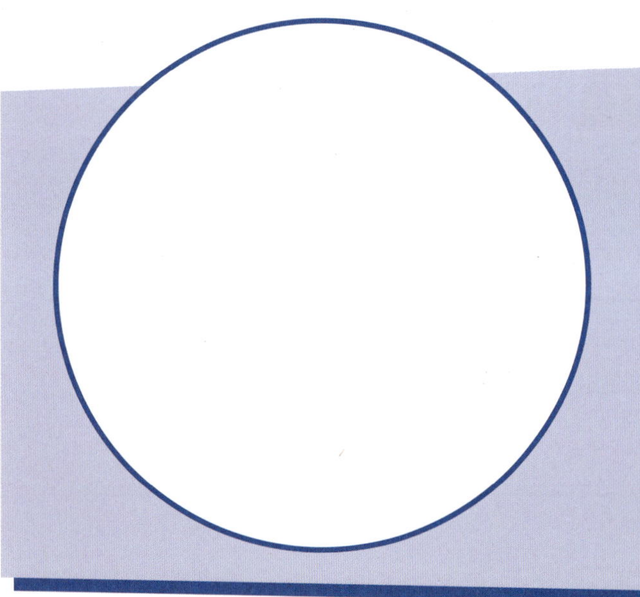

Think about all the ways that you use water at home. Make your own pie chart based on your best guesses of how much water you use for each activity.

What are some things you might do to conserve water? Make a list of the ways you can cut back on water usage.

____

____

____

____

____

Water on Earth

 ISBN: 978125726534

Define It!

balanced: equal forces acting on an object

equal: the same

force: a push or a pull

unbalanced: unequal forces acting on an object

What makes a rocket launch into the sky or a sled speed down a hill? **Forces** make things move. A force is a push or a pull. Forces are everywhere. You cannot see a force, but you can see what it does. A pull opens a door. A push makes a toy car speed away.

Forces cause objects to move, but not always. What would happen if you and your friend each pulled on opposite ends of a rope? If you both tugged with **equal** force, the rope would stay in place. If you used greater force than your friend, the rope would move toward you. When the forces on an object at rest are **balanced** (equal), the object stays at rest. When the forces are **unbalanced** (not equal), the object moves.

Concepts:

Forces make things move.

Forces can be balanced or unbalanced.

Look at each picture. Write *push* or *pull* to describe the force being used.

____

____

____

Friction

Concept:

Friction is a force that slows objects down.

Friction is a force between objects that are touching. When **surfaces** rub together, they grab on to each other. Friction slows down a moving object.

Think of a box sitting on a table. The box is not moving because the forces on it are balanced. If you push it, the forces become unbalanced and the box slides across the table. Will the box slide forever? No, because friction between the box and the table will slow the box down until it stops. Most often, a smooth surface has less friction than a rough surface.

Sometimes friction can be helpful. Friction makes the tyres and brakes stop a bike. Boots with rough soles have more friction to keep you from slipping on an icy sidewalk.

Define It!

friction: a force that slows down the motion of an object that is touching something else as it moves

surface: the outer layer

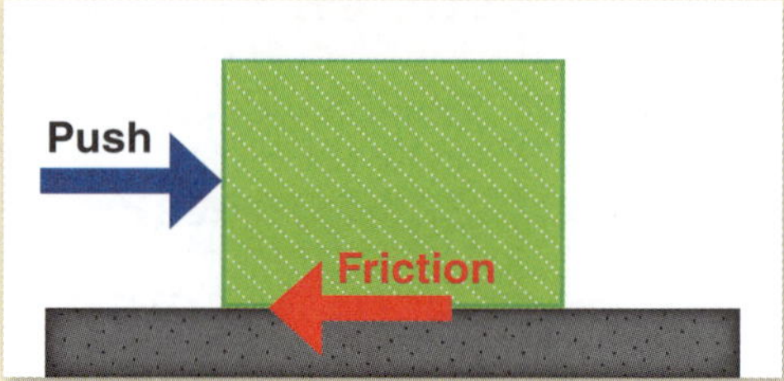

Circle the answer.

1. Which one would work best with greater friction?

tyres and brakes

slide

2. Which one would work best with little friction?

snow boots

sled

TARGETING SCIENCE YEAR 4 © PASCAL PRESS ISBN: 978125726534

Force and Motion

Concepts:

Forces change the motion of objects.

Forces have size and direction.

Define It!

direction: the line along which something travels

motion: the act of moving

size: the amount of something

Forces change the **motion** of objects. Forces make objects speed up and slow down. Forces make objects change direction and change shape.

speed up

slow down

change direction

change shape

A force has both **size** and **direction**. Think about a soccer ball. What happens when you try to make a goal? To get the ball into the goal, you kick it with the right amount (size) of force. If you use too little force, the ball will not reach the goal. You also aim (direct) the force so that the ball goes into the net. If the direction of the force is off, the ball will miss the goal.

Draw a line under the answer.

1. Squeezing a tube of toothpaste shows that force can ____.
   slow things down **change the shape of things**

2. Batting a cricket ball shows that force changes the ____ of an object.
   direction **size**

Forces & Motion

Picturing Forces

Skill:

Infer meaning from information communi-cated in diagrams.

Forces cannot be seen, but we can see what they do. Forces change the motion of objects. In order to show how forces work, scientists make drawings. Arrows in the drawings show the direction of the forces. Arrows also show the size of the forces. A longer arrow stands for a stronger force.

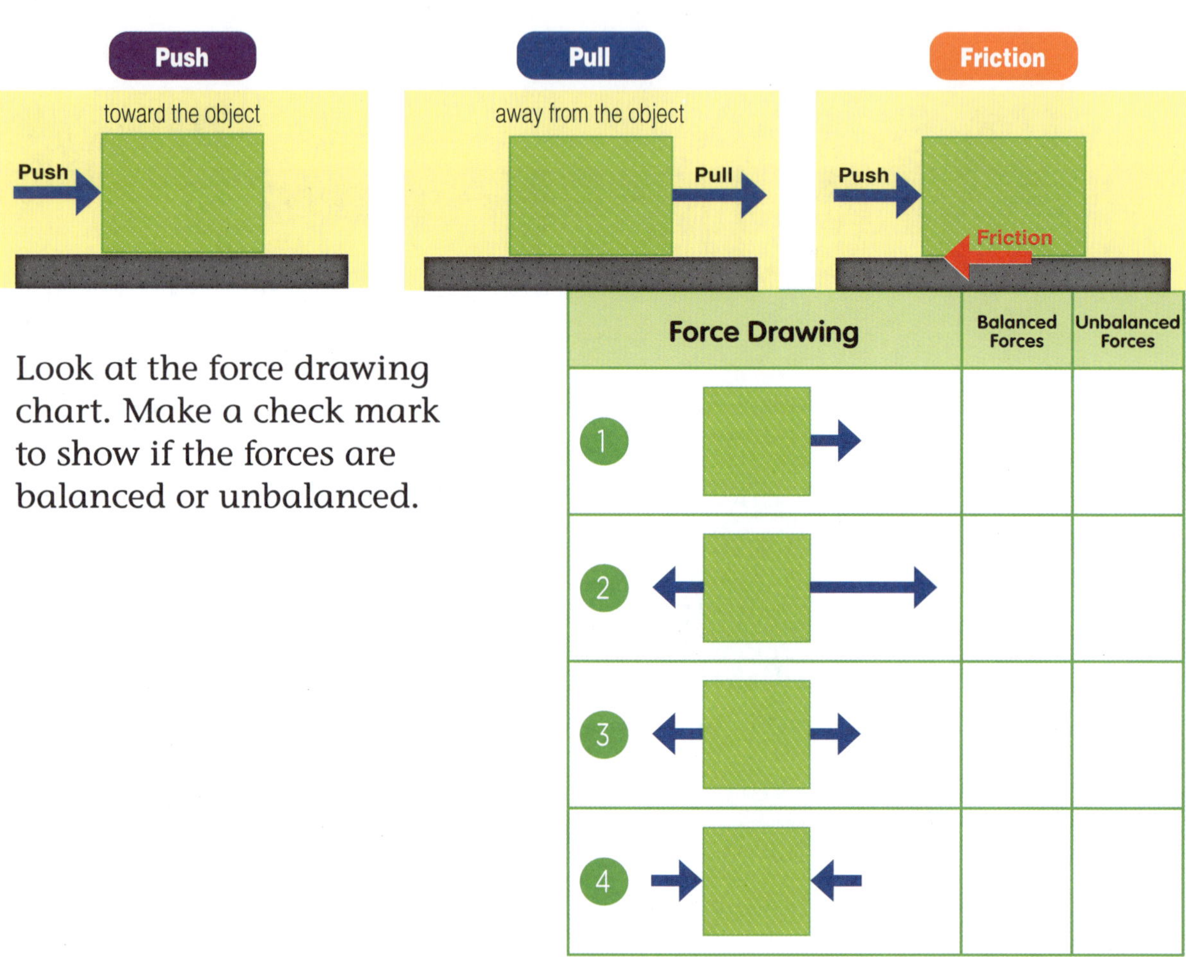

Look at the force drawing chart. Make a check mark to show if the forces are balanced or unbalanced.

Force Drawing	Balanced Forces	Unbalanced Forces
1		
2		
3		
4		

Explain your answer for number 2.

____

____

____

Forces & Motion

TARGETING SCIENCE YEAR 4 © PASCAL PRESS ISBN: 978125726534

Look at this picture. What are the forces acting on this horse and cart?

Gravity is pulling the horse and cart down towards the centre of the Earth.

The surface of the Earth is **pushing** the cart and stopping it from falling to the centre of the Earth.

There is **friction** (a pushing force) between the horse's hooves and the road, the cart wheels and the road, and lots of other places where something is rubbing against something else.

The horse is **pulling** the cart forwards.

We can turn this picture into a **force diagram**, showing where the forces are acting on each element. In force diagrams in physics, we use arrows to show the direction of a force. The longer the arrow, the greater the force.

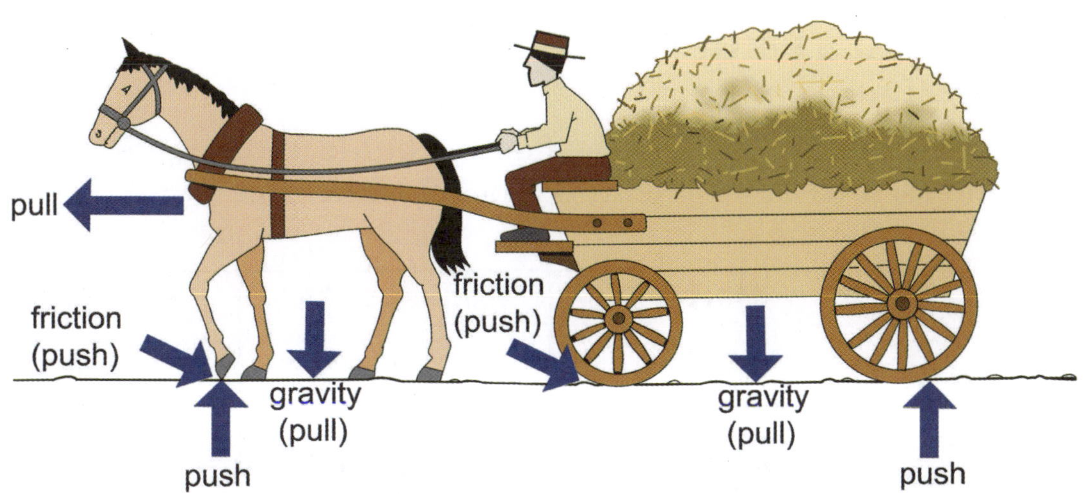

1. Can you label the forces that are acting on this person and barrow? Use a ruler to draw your arrows.

Liquid Forces

Ever wondered why some things float, and other things sink? It's all to do with the forces acting on the object.

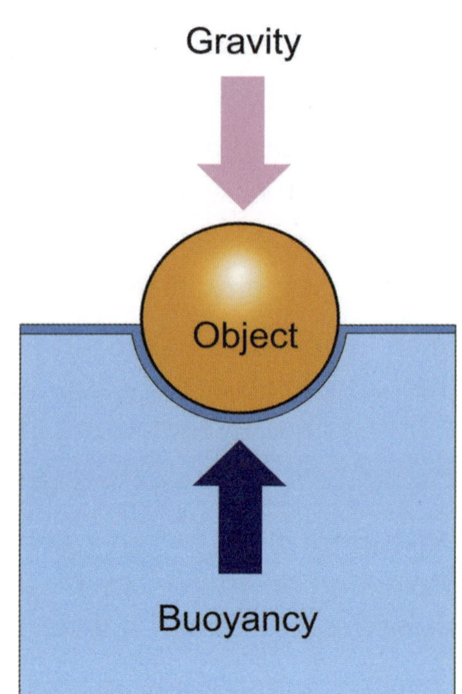

The ball in this diagram is being **pulled downwards** by the force of **gravity**.

The water has what we call **buoyant force**.

This force is able to **push up** against things.

If something floats, the buoyant force is stronger than the gravity of the object.

If something sinks, the gravitational force is stronger than the buoyant force.

So really, it's just a battle of the forces every time.

Which of these things would lose the floating battle and sink, because the gravitational force is stronger than the buoyant force?

cork	Marble	Fishing sinker	Rubber ball
Rock	Rubber duck	Paper boat	Metal Coil
Fork	Leaf	Blob of clay	Paperclip

TARGETING SCIENCE YEAR 4 © PASCAL PRESS ISBN: 978125726534

Forces are always acting between objects. **Contact forces** are when the objects touch each other and **non-contact forces** are when they do not touch each other.

When you kick a ball, the force between the ball and your foot is a **contact force**. The resistance as the ball pushes through the air, and the friction between the ball when it lands on the grass, are all contact forces.

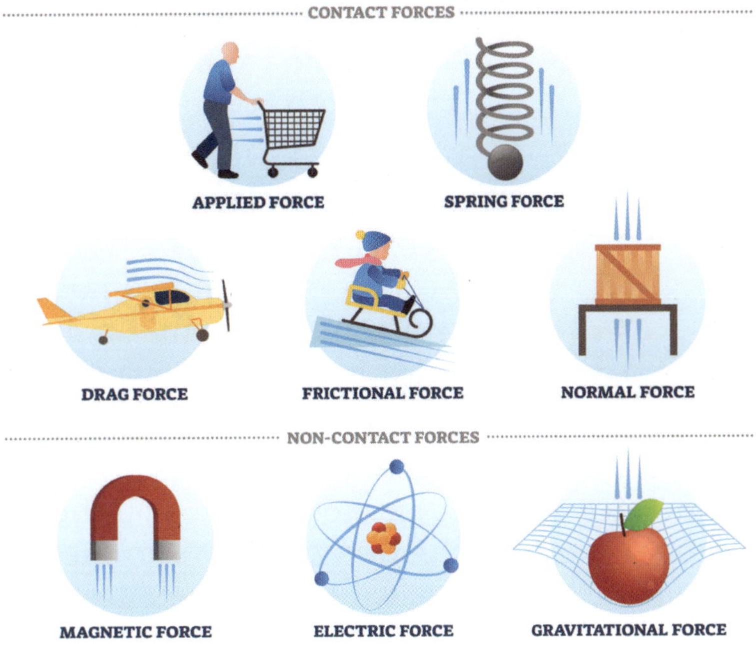

Non-contact forces are harder to see but they are also very important. Magnetic force allows magnets to push and pull each other without touching. If you have ever rubbed a balloon on your head and created static electricity, that's a non-contact force. The biggest non-contact force is gravity. Even if you jump, gravity pulls you back down to Earth. Gravity from the Earth even keeps our moon in place, even though nothing is touching it.

Most of the forces we encounter every day are contact forces.

TYPES OF FORCES

CONTACT FORCES

APPLIED FORCE

SPRING FORCE

DRAG FORCE

FRICTIONAL FORCE

NORMAL FORCE

NON-CONTACT FORCES

MAGNETIC FORCE

ELECTRIC FORCE

GRAVITATIONAL FORCE

Which of these are contact forces, and which are non-contact forces?

Hitting a tennis ball 	Blowing a feather 	A rock dropping to the ground 
Electricity 	A magnet pushing another 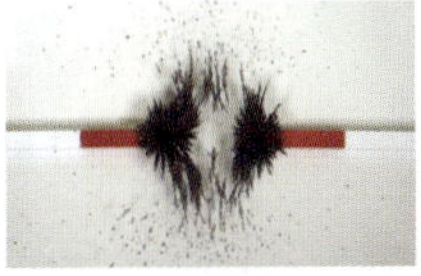	Rubbing your hands together 
A plane's propellors 	Catching a frisbee 	The moon in space 

FALLING ROCKS

First Nations Perspectives: Make a Mammandur

The Mammandur is a spinning top that was played with by First Nations children of Awngthim, Weipa in Queensland. It uses both contact and non-contact forces to work.

Traditionally, the mammandur was made from beeswax and a stick. It was spun by rubbing the stick between two palms or by using the thumb and middle finger to twist it.

You can make your own mammandur by using some blue tack and a round toothpick. Roll the bluetack into a ball and push it onto the stick, about a third of the way from the bottom.

Find a smooth, flat surface and give it a spin!

Think about how your Mammandur moved and the forces acting upon it. Can you place these force words in the correct spots on the diagram and show the direction of the force with arrows?

Gravity
Friction
Push
Pull

TARGETING SCIENCE YEAR 4 © PASCAL PRESS ISBN: 978125726534

Skill:

Apply content vocabulary.

Write each answer.

1. Is a push a size **or** a force? ____________________

2. Is a pull a force **or** a surface? ____________________

3. If forces are balanced, are they equal **or** unequal? ____________________

4. Does friction slow things down **or** speed things up? ____________________

5. If an object moves, is it at rest **or** in motion? ____________________

6. Is the surface of a slide smooth **or** rough? ____________________

7. Does the direction of a force control how far an object goes **or** where an object goes? ____________________

8. Are unbalanced forces equal **or** unequal? ____________________

9. Does the size of a force change the speed of an object **or** the direction of an object? ____________________

10. Is a force seen **or** unseen? ____________________

Think About It

Is a force real or unreal? How do you know?

____

____

TARGETING SCIENCE YEAR 4 © PASCAL PRESS ISBN: 978125726534

Marshmallow Popper

Skill:

Conduct a simple scientific investigation to answer the question "how?"

How far can you make a marshmallow fly? This activity shows how the size of a force changes the distance an object travels.

What You Need

- yoghurt cup
- scissors
- balloon
- mini marshmallows
- pavement chalk

What You Do

1. Ask an adult to cut out the bottom of a yoghurt cup.
2. Tie a knot at the open end of the balloon and cut off about 1 cm from the other end.
3. Stretch the cut end of the balloon over the top rim of the yoghurt cup.
4. Go outside on a paved surface if possible. Drop a mini marshmallow into the cup, pull back on the knotted end of the balloon, aim into the distance, and let go.
5. How far did the marshmallow travel? Mark the place where it landed with pavement chalk. How can you make the distance longer or shorter? Test your idea. Mark each landing place.
6. Pulling back on the balloon and then letting go creates a force that pushes the marshmallow.

How did you make the force greater?

____

How did the size of a force change the distance travelled by the marshmallow?

____

____

TARGETING SCIENCE YEAR 4 © PASCAL PRESS ISBN: 978125726534

Friction in Action

Skills:

Predict the outcome of a simple investigation and compare the result with the prediction.

What You Need

- toy car
- 3 cardboard pieces about 10 x 46 cm
- 3 coverings such as crinkled or smooth foil, bubble wrap, sand-paper, wax paper, paper towel, or other materials
- tape
- 3 stacks of books or a long rectangular object
- measuring tape

What You Do

1. Make three toy car ramps from cardboard. Make them all the same length, about 50 cm long.
2. Cover each ramp with a different covering and tape it. Use both smooth and rough coverings to create different amounts of friction.
3. Prop one end of each ramp on a stack of books or a long rectangular object as shown. If you are using stacks of books, make sure that they are all the same height.
4. Start the car at the top and let it go down the ramp without pushing it. For each ramp, measure the distance the car travels from the bottom edge of the ramp. Write it down.

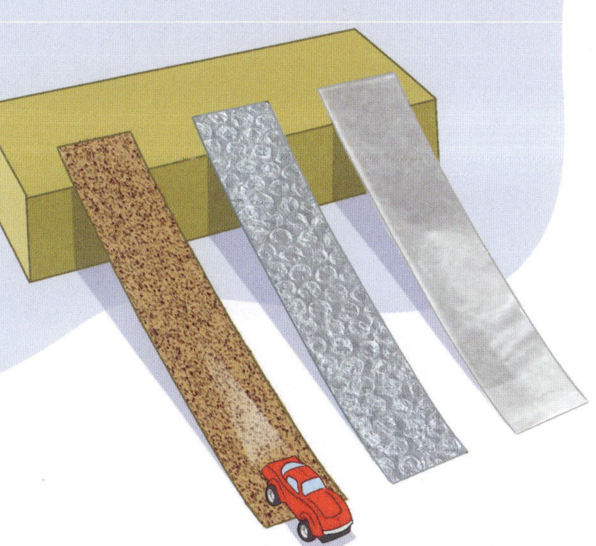

I found out that the ramp covered with ______________ had the least friction.

Forces & Motion

A Law of Energy

Concepts:

Energy is the ability to do work.

Energy can be neither created nor destroyed.

Define It!

conservation: saving the total amount of something

law: something that always happens under the same conditions

transfer: to move from one place to another

universe: all of matter and space

work: the use of force to move an object

Have you ever seen a bolt of lightning flash across the sky? Then you have seen a giant spark of energy. But what is energy? Scientists cannot say exactly what energy is. They do know some things about energy, however. Energy makes it possible to do **work**; that is, to cause something to move. It takes energy for you to run, and it takes energy for a car to move.

Scientists also know that energy can be neither created nor destroyed. This is known as the **law** of **conservation** of energy. If a girl runs across a soccer field and kicks a ball, the energy is **transferred** from the girl's body to the ball. The ball is set in motion. The energy is not lost. Scientists think that the total amount of energy in the entire **universe** always stays the same.

Circle the answers.

1. Energy cannot be transferred. **true** **false**
2. Energy cannot be destroyed. **true** **false**
3. When you kick a ball, the energy transfers to the ball. **true** **false**
4. Scientists know exactly what energy is. **true** **false**

TARGETING SCIENCE YEAR 4 © PASCAL PRESS ISBN: 978125726534

Concepts:

Potential energy is stored energy.

Kinetic energy is the energy of motion.

Faster objects have more kinetic energy.

Define It!

collide: to bump into

kinetic: having to do with motion

potential: the stored energy belonging to something

Scientists describe energy in two ways. **Potential** energy is stored energy. For example, the food you eat for breakfast gives you energy for your day. Energy from the food is stored up in your body. Your body uses the energy when you run a race or give someone a push on a swing at recess. A bowling ball lifted into the air has potential energy and so does a folded-up toy spring.

Kinetic energy is the energy of motion. An object that is moving from one place to another has kinetic energy. Think of a fast-moving car that **collides** with a road sign, knocking it over. The same car moving at a slow speed only bends the sign when they collide. This is because faster objects have more kinetic energy.

STOP

Write the answers.

1. What is another word for stored energy? ________________

2. What is kinetic energy? ________________

3. If Mason throws a fast snowball and Jake throws a slow one, whose snowball has more kinetic energy? ________________

Energy on the Move

Concepts:

Energy can be transferred from place to place.

Energy is transferred through sound, light, and electric current.

Define It!

electric current: the flow of electricity

generate: to make or create

power plant: a factory for generating power

vibrate: to move back and forth quickly

Energy can be transferred from place to place. One way this happens is through sound. A drummer striking a drum uses energy. Striking the drum causes it to move, or **vibrate**. This pushes the air around the drum, making sound waves. The sound waves transfer the sound energy to your ears.

Energy is also transferred by light. Light transfers energy from the sun at a high speed as light moves through empty space. The speed of light is about 300,000 kilometres per second. It only takes about 8 minutes and 20 seconds for light from the sun to reach you on Earth!

Energy can also be transferred by **electric current**. What happens when you plug in and turn on a toaster? Electric current **generated** by a **power plant** moves along wires to your home. The electric current flows to the toaster, supplying the energy the toaster needs to work.

List three ways that energy can be transferred from place to place.

1. ______________________________

2. ______________________________

3. ______________________________

TARGETING SCIENCE YEAR 4 © PASCAL PRESS ISBN: 978125726534

Skills:

Interpret information gained from a graph.

Demonstrate understanding of relative quantities expressed as percentages.

Think of the many things in a home that use some form of energy to make them work. *Appliances* such as refrigerators, washing machines, dryers, and dishwashers use energy. *Electronics* such as computers and televisions use it. Heating systems, air conditioners, and lights all use energy. This graph shows how energy is used in homes.

Read the graph and answer the questions.

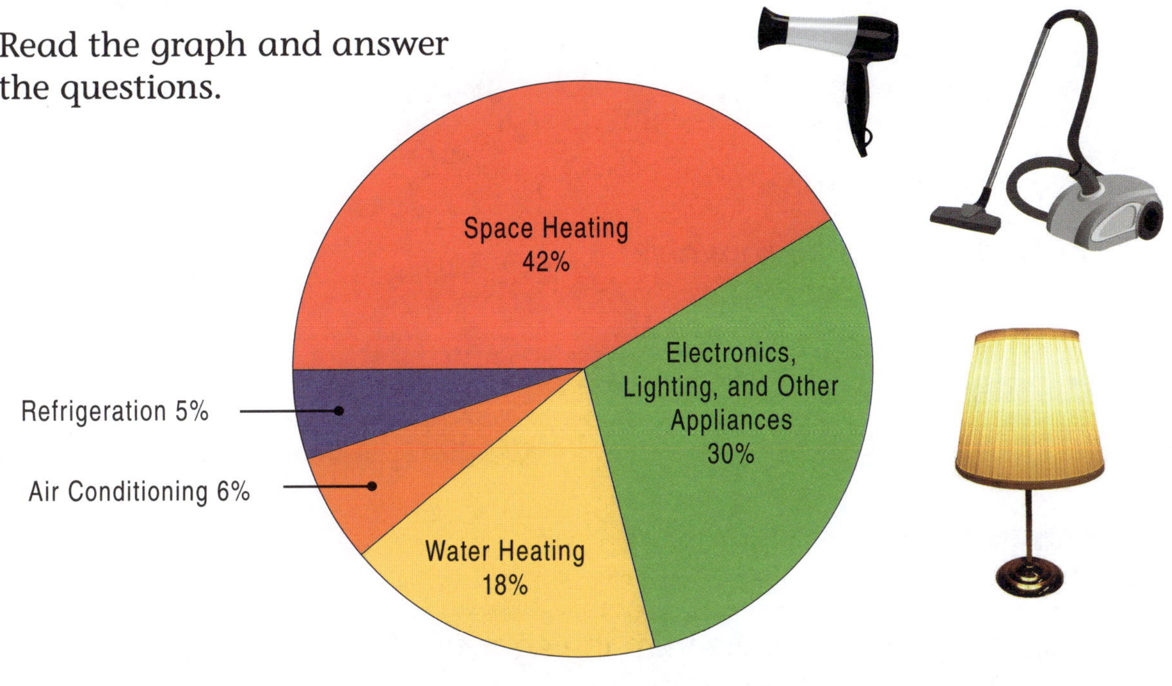

U.S. Energy Information Administration 2009

1. Which is used more—energy for heating or for cooling?

____

2. What is 30% of the energy used for in homes?

____

3. List in order the top three uses of energy in homes.

____

____

____

Can Energy Change Matter?

Skills:

Conduct an investigation by following written instructions.

Record observations and draw conclusions.

Learn about the changes that energy can make with this activity. Observe what happens to the white of an egg when you apply energy.

Note: Be careful when handling raw eggs. Eating uncooked eggs (or licking a beater) can make you sick. When you finish this activity, wash your hands and tools with soap and hot tap water.

What You Need

- white of an egg
- eggbeater or wire whisk
- bowl
- timer

Directions

1. Separate the egg white from the egg yolk. Put the egg white in a bowl. Look closely at the egg white. What colour is it? How much does it fill the bowl? Does it look smooth and shiny? Record what you observe in the chart on page 79.

2. What will happen if you use energy to beat the egg white? Do you think the egg white will still be an egg white, or will it become a different thing? Write your prediction here.

___

___

3. Use an eggbeater to beat the egg. Put lots of energy into it! What happens? Record what you observe.

TARGETING SCIENCE YEAR 4 © PASCAL PRESS ISBN: 978125726534

Time	Appearance of Egg White
Before beating	
After beating	
5 minutes later	
10 minutes later	
15 minutes later	
20 minutes later	
25 minutes later	

4. Do you think it is still an egg white after beating, or has it become a different thing? Give your reasons.

____

____

5. Let the egg white stand for 5 minutes. Then record what you observe in the chart. Do the same thing after 10, 15, 20, and 25 minutes.

6. Do you think that the matter that is left in the bowl after 25 minutes is the same matter you started with? Why?

____

____

Did You Know?

A *physical change* is any change in matter that does not produce new matter. You observed that the energy of beating the egg white changed the look of it, but it was still an egg white. If you repeat the steps, it will still be an egg white. The energy causes a physical change.

Apply What You Learned

Skills:

Analyse and interpret visual images.

Write informative text to explain.

Look closely at the two pictures. Describe what is happening in each picture. Then explain what type of energy is shown.

Hint

Potential energy is stored energy.
Kinetic energy is energy in motion.

____

____

____

____

____

____

Richard Paul Kane / Shutterstock.com

____

____

____

____

____

____

TARGETING SCIENCE YEAR 4 © PASCAL PRESS ISBN: 978125726534

Gravity is a force that pulls objects toward each other. Although you cannot see it, the force of gravity is everywhere. Earth's gravity pulls in one direction—down. A better way of saying that is, gravity pulls you and everything on Earth toward Earth's centre. Gravity is what holds us on Earth.

Isaac Newton was a **scientist** who lived more than 300 years ago. Newton was the first to figure out that there must be a force pulling things to Earth's centre. A famous story says that Newton saw an apple fall from a tree. This made him wonder why things fall down and not sideways or up. He thought there must be a force in Earth's centre that pulls everything toward it. No one knows for sure if the story about the apple is true, but Newton's **law** of gravity was very important to science.

Define It!

gravity: a force that pulls objects together

Isaac Newton: a famous English scientist and mathematician (1642–1727)

law: in science, an observed fact that something always happens under the same conditions

scientist: someone who is an expert in a science

Concepts:

Gravity is a force that holds us on Earth.

Isaac Newton had the idea that there must be a force pulling things to Earth's centre.

Complete the sentences.

1. An apple falls down from a tree, instead of up, because

   ____

2. ____________________ thought about why objects fall to Earth.

Everything Has Gravity

Concepts:

Everything has gravity.

Objects with more mass have a stronger pull of gravity.

Define It!

attract: to pull

mass: how much matter something has

matter: something that takes up space and has weight

Gravity is a force that **attracts**. Everything has gravity, which means everything pulls on everything else. Some objects have a stronger pull than others. Even your body has gravity, but its pull on the things around you is very weak.

The size of the force of gravity depends on the **mass** of an object. An object's mass is how much **matter**, or "stuff," it has. For example, Earth has much more mass than a basketball does, so Earth has a much greater pull. Because of the strong pull of Earth's gravity, the basketball will always fall to Earth no matter how hard it is thrown. And because Earth has much more mass than you do, you will not float off Earth into space.

Circle the answer.

1. Which one has more gravity?

girl Earth

2. Which one has more mass?

book Earth

TARGETING SCIENCE YEAR 4 © PASCAL PRESS ISBN: 978125726534

Define It!

astronaut: a person who travels to outer space

weight: a measure of the pull of gravity on an object

weightless: having no gravity pulling on it

To a scientist, mass and **weight** are not the same thing. Your weight is a measure of the pull of gravity on you. **Astronauts** travelling in space are **weightless** because there is no gravity pulling on them. Their weight is zero. But their mass is the same as it was on Earth.

Other planets, moons, and stars have their own gravity. So if you visited other worlds, your weight would be different from your weight on Earth. A person who weighs 27 kg on Earth would only weigh about 10 kg on Mars. The same person would weigh about 64 kg on Jupiter.

Earth

Jupiter

Earth has a very large mass. This means that it has a strong force of gravity, too. The moon is smaller than Earth and has less gravity. Would you weigh more on the moon or less? Why?

____

____

____

Concept:

Weight is a measure of the pull of gravity on an object.

Playground Forces

Skills:

Explain the relationship between the motion of an object and the pull of gravity.

Infer meaning from data communicated in diagrams.

Gravity

Have you ever looked at a playground to see forces in action? Let's look at what happens when Ava sits on a swing. A force is needed to set the swing in motion. The swing will move when someone pushes Ava or when she pulls back and pumps her legs. After that, forces carry Ava up and gravity pulls her down. You can predict the pattern of motion. If the swing goes forward and up, next it will go down. Then the swing will go backward and up, and down again. When Ava drags her feet on the ground, she creates enough force to stop the swing. If Ava simply stopped pumping her legs, the swing would stop by itself because of the force of the air.

Number these mixed-up pictures from **1** to **5** to show the pattern of motion.

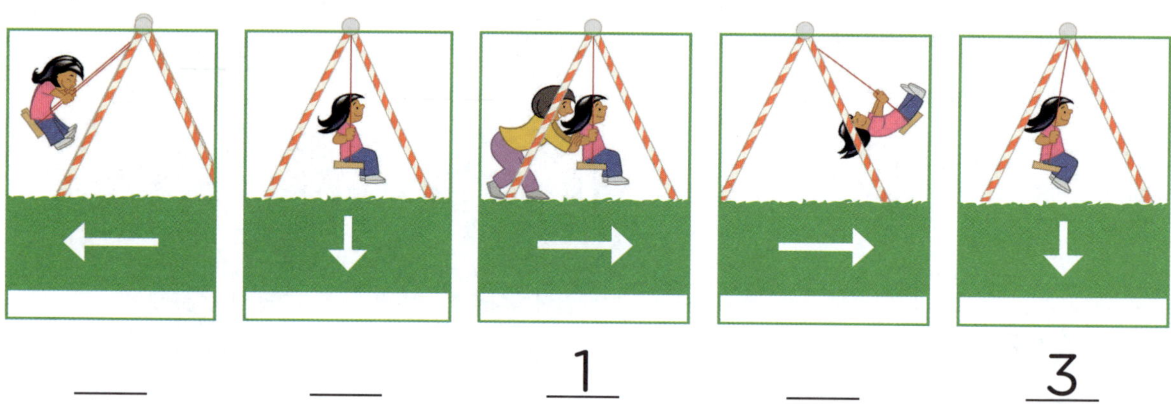

Why does the swing always come down again?

___

___

TARGETING SCIENCE YEAR 4 © PASCAL PRESS ISBN: 978125726534

Gravity Crossword Puzzle

Skill: Apply content vocabulary.

Use the vocabulary words to complete the crossword puzzle.

attract	weight	gravity	Newton
matter	weightless	scientist	astronaut

Across

2. a famous English scientist
5. a force that pulls objects together
6. to pull
8. someone who is an expert in a science

Down

1. a measure of the amount of gravity pulling on an object
3. having no gravity pulling on it
4. something that takes up space and has weight
7. a person who travels to outer space

Falling Water

Skills:

Predict the outcome of a simple investigation and compare the result with the prediction.

Experiment with water to find out more about the effects of gravity on falling objects.

What You Need

- paper cup
- pencil
- water
- grassy area outdoors

What You Do

1. Use a pencil to poke a small hole in the side of the cup near the bottom.
2. Hold your thumb over the hole as you fill the cup with water.
3. Hold the cup up as high as you can reach.
4. Uncover the hole. A stream of water will squirt out of the cup! Gravity is pulling on both the water and the cup, but you are holding the cup to keep it from falling. Only the water can fall. What do you predict will happen to the water if you drop the cup?
5. Test your thinking. Let the cup drop and watch what happens.

What did you observe the water do?

____

____

____

____

Did You Know?

You should have observed the water stop squirting out as the cup fell. That's because gravity was pulling on the cup and the water equally. Even though their masses are different, both the cup and the water were falling at the same rate of speed.

TARGETING SCIENCE YEAR 4 © PASCAL PRESS ISBN: 978125726534

Paper Roller Coaster

Skill:

Follow a sequence of directions to complete a project.

If you have ever ridden a roller coaster, you know the pull of gravity can be fun! Making a roller coaster sculpture is fun, too.

What You Need

- 30 x 46 cm sheet of poster board
- glue and tape
- scissors
- coloured construction paper
- ruler
- crayons or markers
- small piece of tissue paper or other thin paper

What You Do

1. Measure and cut construction paper into long strips 4 to 5 cm wide for the coaster tracks. Decorate the tracks with different patterns.
2. Roll a 15 x 15 cm square of construction paper into a tube to use to hold up the tracks. Tape it. Cut 1.5 cm slits in the bottom of the tube to make tabs for gluing.
3. Let the poster board be your base. Glue the paper tube in one corner of the base. Tape one end of a track to the top of the tube and glue the other end to the poster board.
4. To form the rest of the roller coaster, glue one end of a strip to the poster board. Twist the paper strip and then glue the other end to the poster board. Continue connecting the paper strips in this way.
5. Crumple a paper ball from the thin paper. Send it rolling down different parts of your roller coaster to show the force of gravity.

Gravity

Gravity on the Playground

Skills:

Explain the relationship between the motion of an object and the pull of gravity.

Illustrate the forces acting on objects.

Go to a playground to look for gravity in action. Describe in which direction gravity pulls you. Draw a picture showing the slide, swings, monkey bars, seesaw, or other objects. Then add force arrows to show the pull of gravity.

Draw

Gravity

TARGETING SCIENCE YEAR 4 © PASCAL PRESS ISBN: 978125726534

Define It!

gravity: the natural force of a body that attracts objects to its centre

mass: the amount of matter in an object

orbit: the curved path of an object around a star, planet, or moon

Have you ever heard the saying "What goes up must come down"? This is a true statement everywhere on our planet because of **gravity**. Gravity is a force of attraction that exists between all objects in the universe, including Earth and everything on it. Gravity makes your feet hit the floor after you jump, or a cricket ball fall into a player's glove. And it's the reason we stay on Earth's surface instead of floating off into space.

All objects with **mass** have gravity. The more mass an object has, the stronger its gravitational force will be on other objects. Because Earth is such a massive object, its gravitational force is strong enough to keep us stuck to its surface. Earth's gravity is so powerful, it even keeps the moon in **orbit** around us!

Think about the ways that gravity impacts your life. Name three instances of gravity at work.

1. ______________________________

2. ______________________________

3. ______________________________

Concepts:

Gravity is a force of attraction that exists between all objects in the universe.

The more mass an object has, the stronger its gravitational force will be.

Centre of Mass

Concept:

Gravity pulls toward the centre of Earth.

Define It!

core: the central part of a celestial body

force: a push or pull upon an object; power or energy used on an object

sphere: a round object

Gravity doesn't just pull us down. It pulls us toward Earth's centre of mass. The centre of mass is a single imaginary point inside an object from which gravity seems to act. Because Earth is shaped like a **sphere**, its centre of mass is located in the **core** of the planet.

Gravity is always at work. It pulls everything toward the centre of Earth in a straight line. However, we can use **force** to keep an object from falling or to change its path. For example, when you catch a ball, you use force to stop gravity from pulling it all the way to the ground. And when you swing on a swing, the seat stops you from falling. In addition, the force of pumping your arms and legs makes you swing back and forth. Even when force stops an object from falling or changes its path, gravity is always pulling on it.

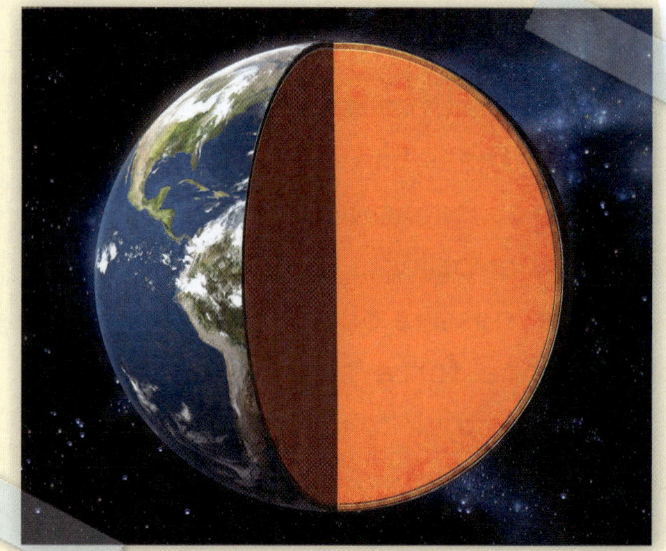

Complete the sentences.

1. You can use ________________ to keep an object from falling or to change its path.
2. Gravity pulls all objects toward Earth's ________________ of mass.
3. Because Earth is shaped like a ________________, gravity acts from the planet's core.

TARGETING SCIENCE YEAR 4 © PASCAL PRESS ISBN: 978125726534

Equal Force

Concepts:

All objects have grav-ity.

Gravity pulls on all objects, no matter what their size, at the same speed.

Define It!

air resistance: the force that opposes the motion of an object through air

encounter: to come upon or meet with something

exert: to apply or put forth

All objects, no matter how small, have gravity. So while Earth's gravity pulls us to the centre of the planet, we all pull on Earth with our gravity, too. The amount of force that gravity **exerts** on an object depends on the mass of the object. The greater the mass of an object, the stronger its pull of gravity is.

Yet gravity pulls all objects—no matter their size or mass—to Earth at the same speed. If you drop a textbook and an eraser from your desk, they will both hit the ground at the same time. However, there are other forces besides gravity that impact the speed at which things fall. All falling objects **encounter** a force called **air resistance**. Air resistance can slow down an object's fall. Because of air resistance, a feather will not hit the ground at the same time as a textbook.

Sky divers use parachutes to increase air resistance and slow down their fall.

Write *true* or *false*.

1. Gravity pulls a textbook and an eraser to the floor at the same speed. ____________
2. The lesser the mass of an object, the weaker its pull of gravity is. ____________
3. Air resistance makes objects fall faster. ____________
4. All objects have gravity. ____________

The Effects of Gravity

Skill:

Label images that represent scientific concepts.

Look at the two sets of images below. Follow the directions to show how gravity is affecting the objects. Then answer the questions.

Draw arrows to show which way gravity pulls on each object.

1. In which direction is gravity pulling the objects? ______________________

2. Which object will hit the ground first? Explain your answer.

____

Draw arrows to show how gravity pulls on the swing.

3. In which direction is gravity pulling the girl in the second picture? ______________________

4. What stops the girl from falling to the ground in the first picture? ______________________

5. What helps change the path of the girl in the second picture?

____

Gravity

 ISBN: 978125726534

Skill: Apply content vocabulary.

Select from the vocabulary words to complete the sentences. Then unscramble the shaded letters to decode the secret message.

gravity	force	core	encounter	exert
air resistance	sphere	mass	orbit	

1. Earth's gravity keeps the moon in ___ ___ ___ ___ ___ around us.

2. Because Earth is shaped like a ___ ___ ___ ___ ___ ___, the centre of mass is in the planet's ___ ___ ___ ___.

3. ___ ___ ___ ___ ___ ___ ___ ___ ___ ___ ___ ___ ___ can slow the speed of an object's fall.

4. Gravity will ___ ___ ___ ___ ___ more force on an object that has more ___ ___ ___ ___.

5. ___ ___ ___ ___ ___ ___ ___ pulls us and all other objects toward Earth's centre.

6. We constantly ___ ___ ___ ___ ___ ___ ___ ___ ___ opposing forces here on Earth.

Even though our mass doesn't change, our weight will be different on another planet because of its gravity. If a person weighs 67.5 kg on Earth, he will weigh only 26 kg on ___ ___ ___ ___ ___ ___ ___.

TARGETING SCIENCE YEAR 4 © PASCAL PRESS ISBN: 978125726534

Gravity Pull

Skills:

Conduct experiments, record data, and analyse results.

The famous thinker Galileo argued that two objects of different mass that are dropped at the same moment from the same height should hit the ground at the same time. Legend says he dropped items from the Leaning Tower of Pisa to prove the point. The following experiment will demonstrate why his argument was correct.

What You Need

- ball
- small rock
- feather
- wad of cot-ton
- metre ruler

Directions

1. Line up the ball, small rock, feather, and wad of cotton at the edge of a desk or table.
2. Use the metre ruler to push all the objects off the table at the same time. You may also ask a friend or parent to push them at the same time as you. You want to be sure the objects are falling at exactly the same moment.
3. Watch to see which objects hit the floor first. Also, listen for the sound each object makes when it hits the floor.
4. Record the order in which the objects hit the floor on page 95.
5. Repeat the experiment as often as needed to be sure of your results.

TARGETING SCIENCE YEAR 4 © PASCAL PRESS ISBN: 978125726534

Record your observations in the table.
Then answer the questions below.

	Order objects hit the floor			
	Test 1	Test 2	Test 3	Test 4
Ball				
Small rock				
Feather				
Wad of cotton				

What Did You Discover?

1. Which object(s) hit the floor first? ____________________

2. Which object(s) hit the floor last? ____________________

3. Which objects do you think encountered air resistance? How do you know?

____

____

4. What do you think would happen if you dropped the items in a vacuum, where there is no air?

____

____

A World Without Gravity

Skill:

Write narratives to develop real or imagined experiences or events.

Imagine what would happen if gravity slowly started disappearing from Earth. Write a short story describing which things get impacted first and how you survive in a world without gravity.

Magnets Attract Metals

Concept:

Magnets are attracted to metals with iron.

Define It!

attract: to pull

iron: a blue-grey metal that can be made into a magnet

magnet: an object that attracts iron

magnetism: the force that attracts iron

What makes the notes stick on this refrigerator door? Not glue, but an unseen force called **magnetism**!

Magnetism is the force that makes **magnets** pull, or **attract**, some kinds of metal. The metal refrigerator door is attracted to magnets. A magnet will not attract glass, plastic, wood, or anything else that does not contain metal.

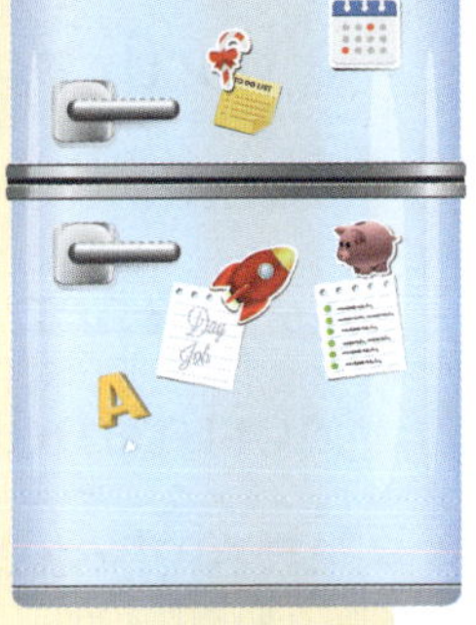

Magnets stick to objects made of metal, but not all metals. The metal **iron**, or a metal that has iron in it, is attracted to a magnet. A magnet won't stick to coins or soft drink cans. They are not made of iron.

1. Draw an **X** on the objects that will **not** be attracted to a magnet.

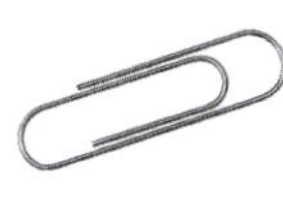

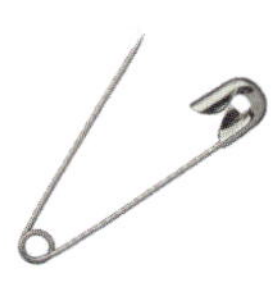

2. Which objects above may have iron in them? How could you test them?

____

____

____

Magnets

Magnets Push and Pull

Concept:

Magnets attract and repel each other.

Define It!

magnetic field: the space around a magnet where its force can be found

pole: either end of a magnet

repel: to push away

All magnets have a **magnetic field** that you cannot see. It's the area around the magnet where a force pulls objects toward the magnet. Magnets have two **poles**. The poles are the parts of the magnet where its force is the strongest. Every magnet has a north pole and a south pole. **N** stands for *north* and **S** stands for *south*.

When two magnets are held with their north and south poles together, the poles attract each other. They pull together. But two poles of the same kind **repel** one another. They push away.

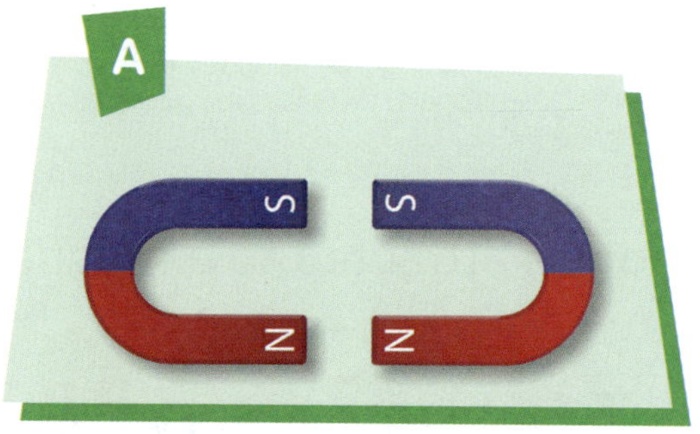

1. Which picture shows magnets that will attract? **A** **B**
2. Which picture shows magnets that will repel? **A** **B**

TARGETING SCIENCE YEAR 4 © PASCAL PRESS ISBN: 978125726534

Magnets: Strong or Weak?

Define It!

magnetic: able to attract iron or act like a magnet

strength: power

weak: lacking strength

Which magnets are stronger—big ones or small ones? You cannot know the strength of a magnet just by its size. The **strength** of a magnet has to do with its magnetic field. A strong **magnetic** field will attract more than a **weak** magnetic field will. The stronger magnetic field will attract things that are farther away or that are heavier.

Look at these two magnets. You cannot see the magnetic force around a magnet. But the lines in these pictures show where the magnetic fields are. The field shown with more lines is stronger. Trace the magnetic fields.

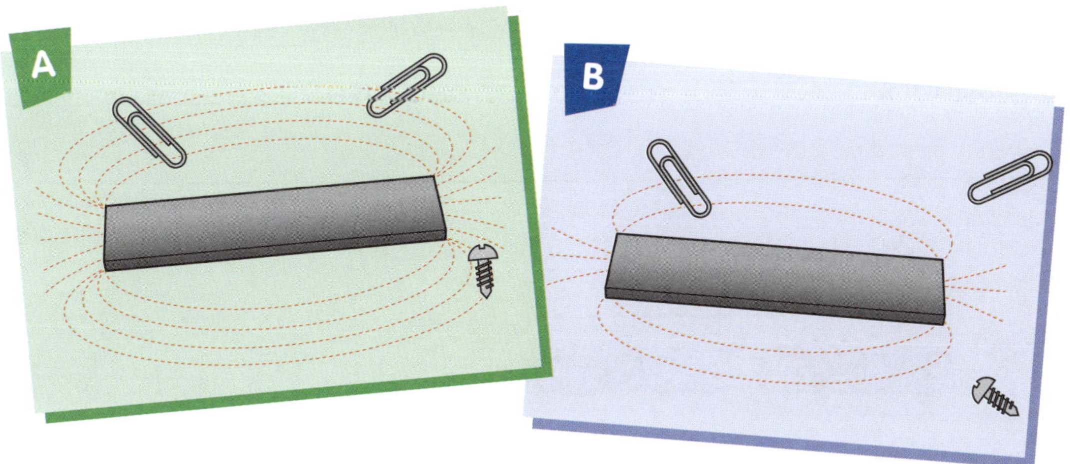

1. Which magnet is stronger? A B
2. Which magnet will attract the most objects? A B

Looking at Magnets

Skill:

Infer meaning from information communicated in diagrams.

Magnets **attract** when the north and south poles are put together. Magnets **repel** when two poles that are alike are put together.

Trace the arrows that show the magnetic forces in each picture. Complete the sentence to tell what is happening in each picture and why. Use the words *attract* and *repel*.

1

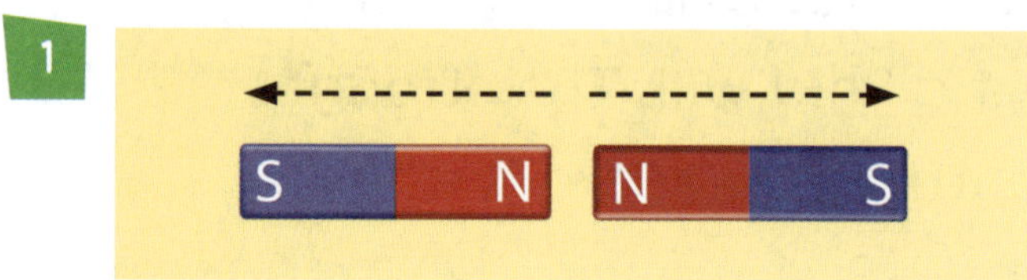

The magnets ____________ each other because ______________________

____

2

The magnets ____________ each other because ______________________

____

3

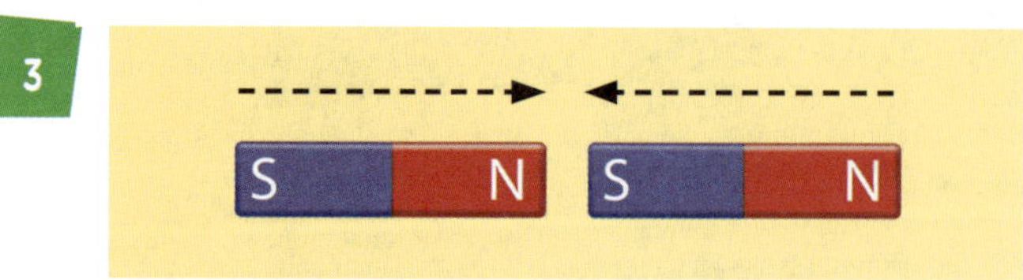

The magnets ____________ each other because ______________________

____

Magnets

TARGETING SCIENCE YEAR 4 © PASCAL PRESS ISBN: 9781925726534

Skill: Apply content vocabulary in context sentences.

Read each clue and write the missing word on the lines.

1. The space around a magnet that attracts metal is its ___ ___ ___ ___ ___.
2. A magnet attracts metal with ___ ___ ___ ___ in it.
3. Another word that means "to push away" is ___ ___ ___ ___ ___.
4. A magnet has two ___ ___ ___ ___ ___: north and south.
5. Magnets do not attract objects made of ___ ___ ___ ___.

Write the letters from the yellow boxes to answer the riddle.

Science Riddle

You cannot see it, but you can see what it does. What is it?

It is a magnetic ___ ___ ___ ___ ___.

Magnet Scavenger Hunt

Skills:

Conduct a simple scientific investigation and record results.

Go on a hunt around your house. First, find a magnet. Then look for items that will stick to the magnet. Which things will **not** stick? Complete this chart. Draw and label six things you found.

Magnetic	Not Magnetic

What shape is your magnet? Draw it here.

Did You Know?

Magnets come in different sizes and shapes, but they all have a north and a south pole. On a bar magnet or a horseshoe magnet, one end is the north pole and the other end is the south pole. If your magnet is a disc, one side is the north pole. Flip the magnet over, and the other side is the south pole.

TARGETING SCIENCE YEAR 4 © PASCAL PRESS ISBN: 978125726534

Make Bottle Cap Magnets

Skill:

Conduct a science experiment.

What You Need

- bottle caps
- patterned paper
- clear packing tape
- gems, beads, buttons
- self-adhesive disc magnets (available in craft stores)
- scissors
- glue

What You Do

1. Cut a circle from patterned paper to fit inside the bottle cap.
2. Protect the paper circle by placing it between two pieces of clear packing tape. Then trim the circle and glue it to the inside of the bottle cap.
3. You may wish to glue gems, beads, or buttons inside the bottle cap to decorate it.
4. Peel and stick a magnet to the back.
5. Go around the house and discover where your magnetic bottle cap will stick!

Note: Do **not** stick your magnet on a computer, monitor, watch, or hearing aid.

Apply What You Learned

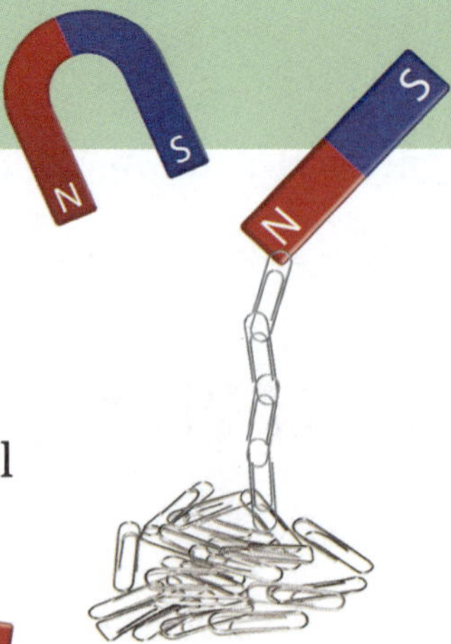

Skills:

Conduct a simple scientific investigation and record results.

Test the strength of two magnets to find out which magnet is stronger. Use two magnets and a handful of paper clips. Write about your investigation and draw sketches to show what you did.

Hint

To test the strength of two magnets, find out how much each magnet can pick up at one time.

1. **Define the Problem:** What do you want to find out?

____

2. **Design a Test:** What will you do? _______________________

____

____

3. **Record Your Data:** Write what you saw when you did the test.

____

____

____

____

____

Draw what you saw.

4. **Write About the Results:** What did you find out?

____

TARGETING SCIENCE YEAR 4 © PASCAL PRESS ISBN: 978125726534

Property Terminlogy

Have a look around you and pick up the nearest thing you can find. Do you know what it is made from? Is it made from one or more type of material? Why do you think that material was chosen to make that particular thing?

Every thing around us has **properties.**

A **property** is a distinguishing quality of an object. For example, its colour, shape, size, temperature or weight.

https://clickv.ie/w/D0gx

Use this QR code to access a video on this topic.

How many of these properties have you heard of before?

Smooth Without bumps or roughness	**Rigid** You cannot bend it without breaking	**Opaque** You cannot see through it
Rough An uneven or irregular surface	**Flexible** You can bend it easily	**Light** Easy to lift
Hard Solid, firm, and rigid; not easily broken, bent, or pierced	**Shiny** Polished so that it reflects light	**Heavy** A lot of weight, hard to lift
Soft Easy to mould, cut, compress, or fold; not hard or firm to the touch	**Dull** Does not reflect light, not shiny	**Brittle** Hard but easily broken
Stretchy Can be pulled to make it longer	**Translucent** You can see through it, but not clearly	**Strong** Able to withstand force, pressure or wear
Elastic When pulled, it returns to its original size	**Transparent** You can see through it clearly	**Weak** Easily broken or torn

Here is a challenge for you:

1. Can you find something that has **one** of these properties? ____________
2. Can you find something that has **two** of these properties? ____________
3. Can you find something that has **three** of these properties? ____________
4. Can you find something that has **four** of these properties? ____________
5. Can you find something that has **five** of these properties? ____________
6. How high can you go? __________ What was the object and what were its properties? ______________________________

Grouping Properties

Look at this box of things you might find in a classroom:

Can you sort them into the boxes based on their properties? You might need to write them in more than one box! An example has been done for you.

Rigid things: • Scissors • Magnifier • Pencil • Ruler	Soft things:	Wooden things:
Flexible things:	Round things:	Hard things:
Opaque things:	Stretchy things:	Smooth things:

TARGETING SCIENCE YEAR 4 © PASCAL PRESS ISBN: 978125726534

Many objects need a combination of materials, with a combination of properties, to make them suitable for their purpose.

This glue stick has to have a rigid barrel to store the glue in. It has to be smooth so that it does not hurt your fingers, cylindrical so that the glue can wind up in it, and light so you can carry it easily.

The glue inside has to be sticky and soft to be useful for paper.

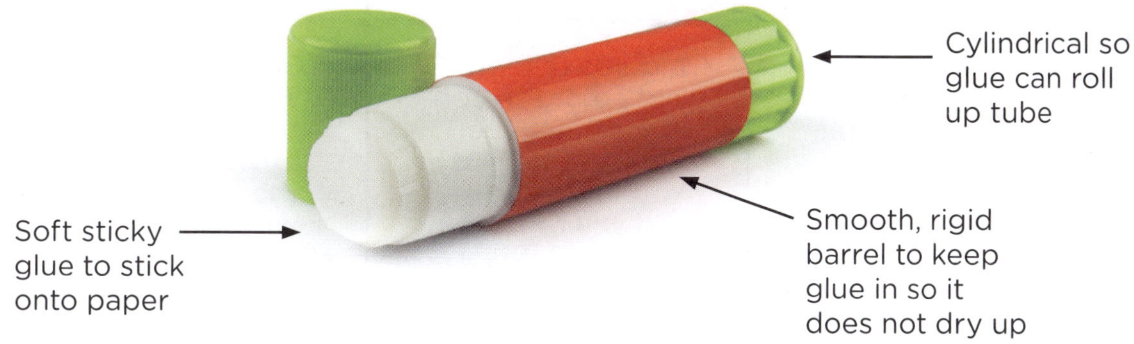

Look at the objects below and explain why each material has been used under the labels.

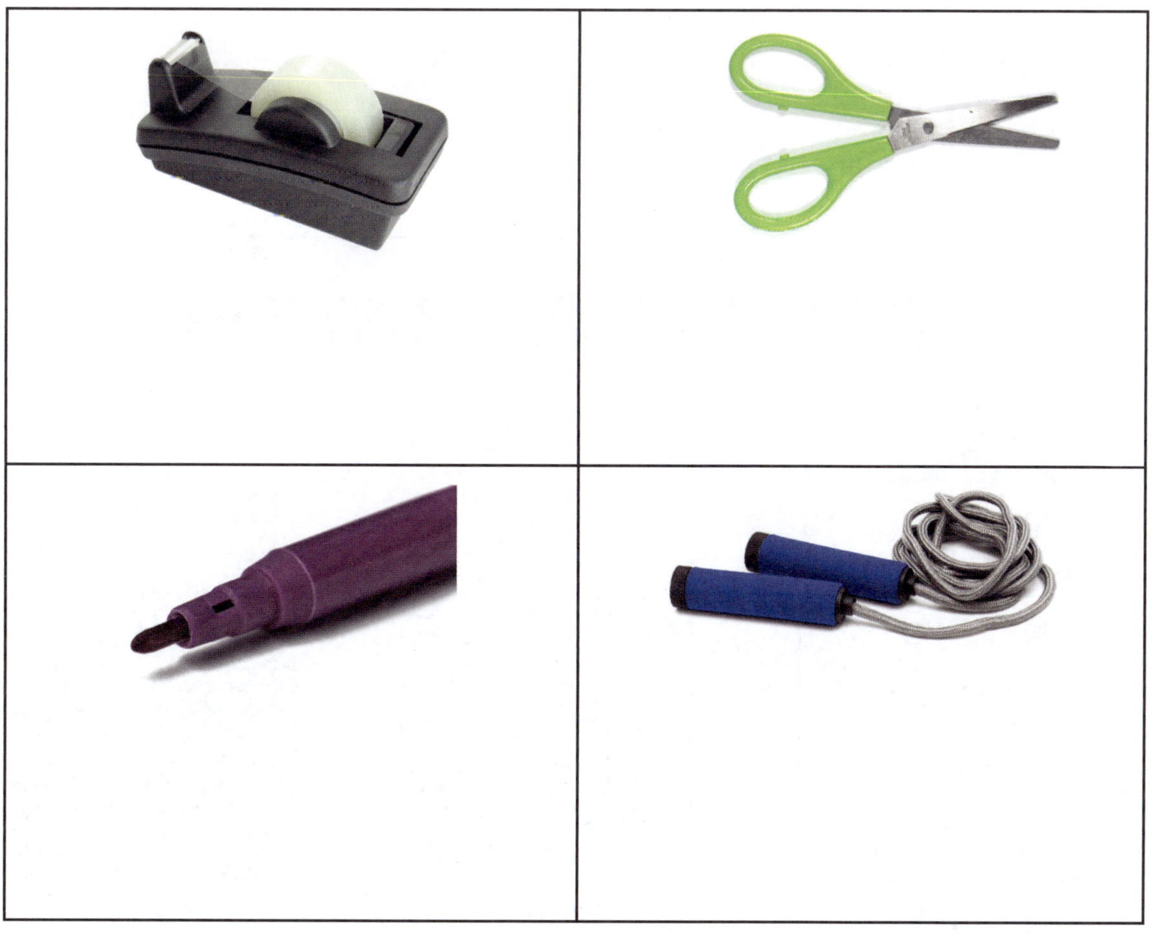

What is matter?

Matter is the 'stuff' which everything is made of. It is anything that takes up space, or can be weighed. Matter is made up of atoms and molecules.
Your teeth, the clothes you are wearing, the seat you are sitting on and the air you are breathing ... it is all matter.
Matter can take one of four states: solid, liquid, gas or plasma. In this unit, we will be focusing on solids, liquids and gases.
Matter can be classified in these groups by the properties it demonstrates. If you can learn to understand these properties and what they mean, it will help you to be able to classify matter into the correct group.

Solids	Liquids	Gases
• Solids hold their shape (they stay the same if you walk away and come back - even over a very long period of time). This is because the molecules which they are made up of are tightly packed together. • Solids do not flow, even if left for a long period of time. • Solids form a mound (or pile) when you pour them, like sugar on a plate.	• Liquids take the shape of the container they are in. If the bowl is round, the liquid will take that shape. This is because the molecules in liquids are not as tightly packed together as they are in solids, so they can move around more within the liquid. • Liquids flow, but some of them flow VERY slowly. • Liquids stay level as they fill a container.	• Gases take the shape of the container they are in, but if you take the lid off, they will also flow out of the container. This is because the molecules in gases can flow freely, and will spread all through the space they are in.

DENSITY OF MATTER

GAS

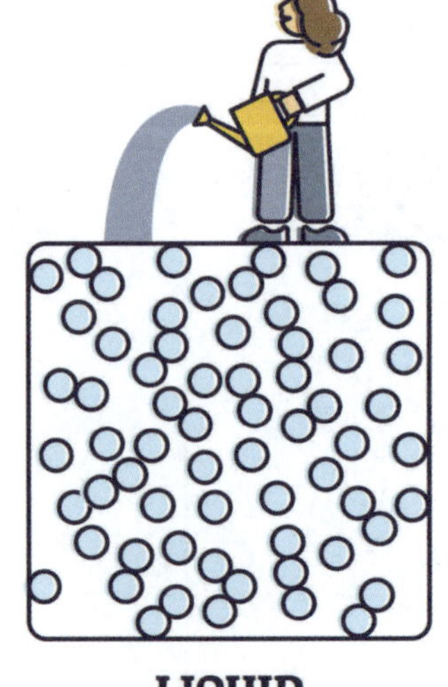

LIQUID

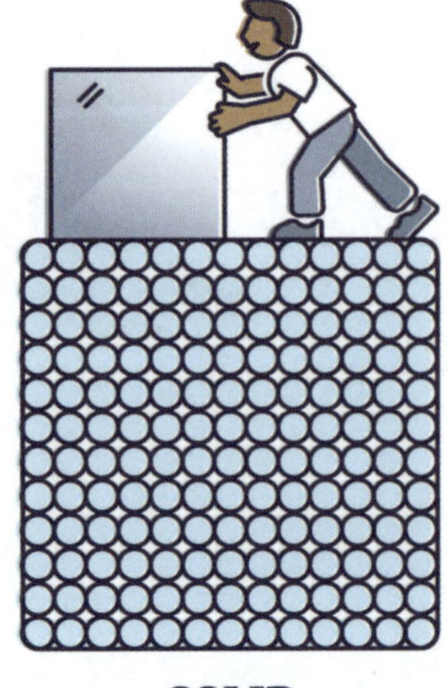

SOLID

TARGETING SCIENCE YEAR 4 © PASCAL PRESS ISBN: 978125726534

Using the guide on page 106, can you classify these substances into solids, liquids and gases? Write what they are in each box under the picture.

The Building Blocks of Matter

An atom is the smallest part of matter. It is the basic 'building block' for all of the matter in the universe. Atoms are very, very small. You could fit about 10 million hydrogen atoms just into the head of a pin!

An element is a substance that is made up of only one kind of atom. Some examples of elements are helium, carbon and oxygen. There are 118 different elements that have been recognised on Earth, and all matter is made of them in some way.

Periodic Table of the Elements

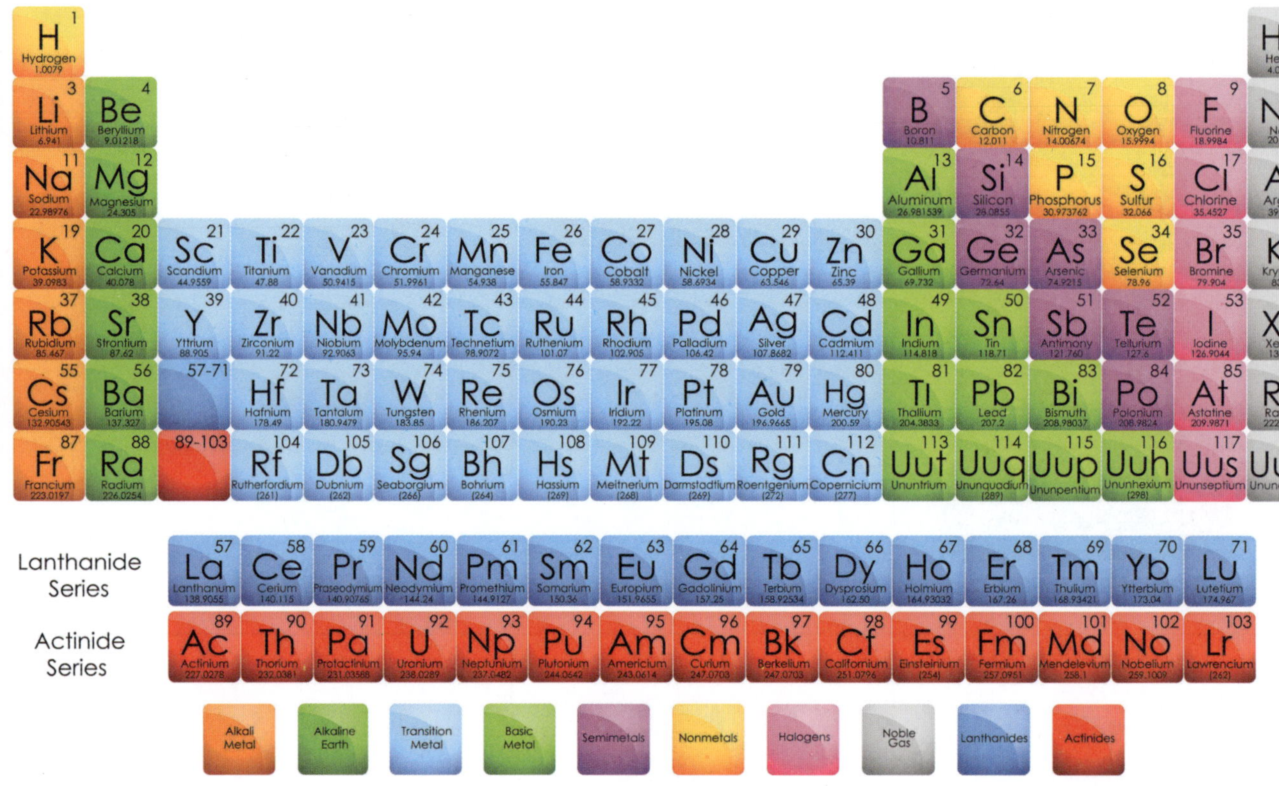

A molecule is made up of either one kind or more than one kind of atom. A molecule of water has two atoms of hydrogen, and one atom of oxygen. The gas oxygen is made up of two oxygen atoms.

TARGETING SCIENCE YEAR 4 © PASCAL PRESS ISBN: 978125726534

In chemistry, most changes to materials can be broken into two groups: Physical Change and Chemical Change.

In a physical change, the material might change shape, size or the form it takes, but no new substance is made.

For example, you could cut a piece of fabric or a block of wood, and it would look different, but you have not made a new substance. It is a change in form, but not in composition.

You could melt an ice block so that it is now a liquid, but you have not made a new substance, just the form that it took.

Some physical changes can be reversed or undone, like re-freezing the ice, but it would be hard to make the wood pieces into a block again, even though it is still wood.

Dissolving sugar in water is another physical change. You have made sugary water, but if you heat it up and the water is evaporated, the water and sugar can be completely separated again, and no new substance has been made.

In a chemical change, a new substance is formed and it can never be reversed, or undone.

For example, if you burned a log, a new substance would be made - charcoal, and you could never make it go back to the way it was before.

You can usually tell if a chemical change is taking place because it often produces heat, or changes in colour, or has an odour or smell.

Are the changes below physical or chemical?

Slicing a tomato		Tearing paper		Cooking a cake	
Defrosting peas		BBQ steak		Freezing water	
Cutting wood		Scrunching tissues		Dissolving sugar	
Frying bacon		Melting plastic		Boiling eggs	

Dissolving Salt - A physical or chemical change?

For this investigation you will need:

- 4 tablespoons of cooking salt (or any other kind of salt)
- ½ a cup of water (warm is better)
- A teaspoon for mixing
- A cup to mix in
- Some black or dark cardboard
- A sunny spot
- An eyedropper or pipette

Procedure:

1. Add the salt water to the water in the cup and mix until it has dissolved, or almost completely dissolved. If it won't dissolve fully, don't worry, it might be 'saturated' which is the scientific way to describe when a solution can't dissolve any more solid into a liquid.
2. Lie the black cardboard or paper in the sun, where it will be able to lie still without blowing away. You might need to put some rocks in the corners.
3. Use the dropper or pipette to suck up the salty solution, and then make a pattern or picture on the card by dripping it on carefully.
4. Leave it in the sun until the liquid has evaporated.

Results:

1. What did you find?

____

____

2. Draw what your solution looks like now.

3. Why did this happen?

____

____

4. Was this a physical or chemical change? How do you know?

____

____

____

Properties of Materials

TARGETING SCIENCE YEAR 4 © PASCAL PRESS ISBN: 978125726534

Comparing Absorbency

In Chemistry, absorbency relates to how well an object can soak up, or take in other substances, usually liquids. Some materials are a lot more absorbent than others, but can you tell by just looking at them?

In this experiment, you are going to predict, test and compare the absorbency of a number of different materials found around most people's homes and schools.

You will need this equipment:

- A piece of paper
- A piece of paper towel
- A piece of toilet paper
- A piece of brown paper or a lunch bag
- A piece of fabric that you are allowed to cut up
- A piece of aluminium foil
- A pair of scissors
- A ruler
- A pen or marker
- An eyedropper or pipette or the smallest spoon you can find

Procedure:

1. Use the ruler and pen to measure and cut a square 10 cm X 10 cm from each material. They need to all be the same size for this to be a fair test.
2. Lie them all beside each other so that you can look closely at all of them. Put them in the order of the material you think will be the most absorbent to the least absorbent before you test them.
3. One at a time you are going to put the exact same amount of water in the middle of each material, then watch it closely to see how the water acts.
4. Does the water sit on top, or soak through? Does the water spread out across the material, or go through and leave a large wet patch underneath when you lift it up? Write your observations in the box below as you test each one:

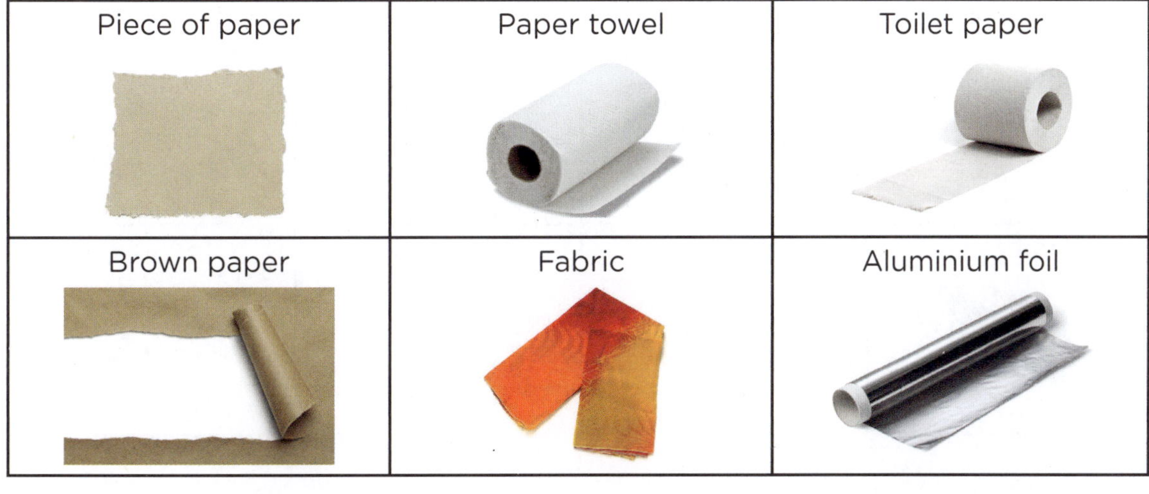

Piece of paper	Paper towel	Toilet paper
Brown paper	**Fabric**	**Aluminium foil**

TARGETING SCIENCE YEAR 4 © PASCAL PRESS ISBN: 978125726534

Comparing Absorbency

Results:

1. Write the numbers 1 to 6 in the corner of the boxes where you made your observations. Write 1 for the material that was the most absorbent and 6 for the one that was the least absorbent.

2. What did you notice for the most absorbent material? How did you know it was absorbent?

____

____

3. The foil did not absorb the water at all. When something does this, we say it is water repellent, or waterproof. Can you think why you would want foil to be waterproof?

____

____

4. What other things in your home or school are waterproof?

____

____

5. Look at the pictures of the things below. Are they water-proof or absorbent?

Towel	Plastic dish	Toilet paper	Beach ball
Sponge	Nappy	Bread board	Puppy pad
Glass	Cereal bowl	Swimming pool	Sand pit
Umbrella	Coffee mug	Paper towel	Tea towel

 TARGETING SCIENCE YEAR 4 © PASCAL PRESS ISBN: 978125726534

Natural and Man-made Materials

All of the things that we use every day have to come from somewhere on Earth. Some of them, like rocks, can be used as they are found. These are called natural materials. Other materials need to be changed by people to be useable, like trees being made into paper. These are called man-made materials.

Some natural materials are found under the ground, like coal and rock, whereas others come from living things on the Earth's surface.

Natural Materials from under the ground:		**Natural Materials from living things:**	
• Stone	• Iron	• Honey	• Leather
• Rock	• Coal	• Wood	• Natural Rubber
• Sand	• Clay	• Wool	• Silk
• Chalk	• Metal	• Cotton	• Organic oil

Man-made materials start from products found on above or below the ground, but then have to be changed in some way to make them more useable:

Man-made materials:			
• Concrete • Vinyl • Glass	• Nylon • Paper • Plastic	• Synthetic rubber • Steel	• Rayon • polyester

Can you work out where these natural things come from? Match the pictures to each other with a line:

Choosing Materials

Imagine you are in charge of building and decorating your very own house. What materials would you use both in the building and the decorations? Draw an annotated diagram of your house and label the materials you would use, and whether these materials are natural or man-made. Your house does not have to be as fancy as this one!

TARGETING SCIENCE YEAR 4 © PASCAL PRESS ISBN: 978125726534

Look at these home-made bird feeders.

1. What is their purpose?

____

2. What properties do they have in common?

____

3. What properties are different?

____

4. What things would you need to consider if you were building a bird feeder for the birds in your area?

____

____

____

5. Draw a plan of a bird feeder you would like to make. Make sure you label the materials you would like to use, and explain why the properties of those things are important for your feeder.

Recycling

https://clickv.ie/w/Lzgx

Use this QR code to access a video on this topic.

In Australia, these things can be recycled:

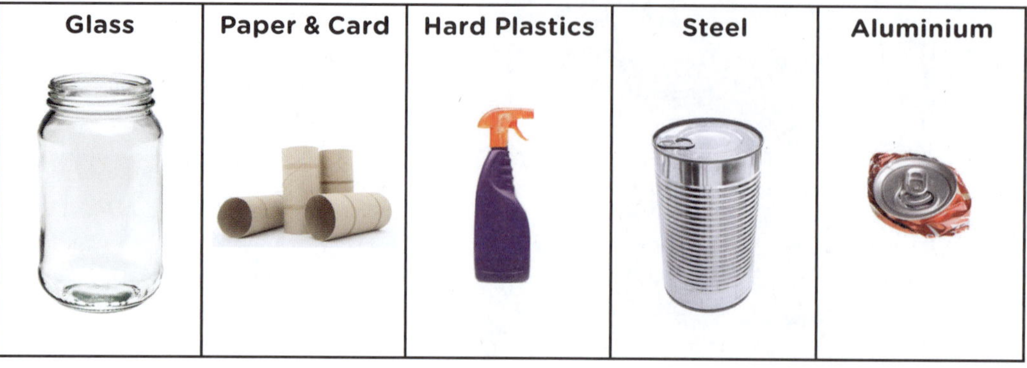

What is a single-use plastic? Have you ever heard that term before?

Single use plastics are the ones we throw out after we have just used them one time. The soft plastics (like the cling wrap on your lunch) are tricky to recycle (they can't go in the normal recycling bin) and, if not recycled, the hard plastics can take decades, and even centuries to decompose.

In Australia, we throw out about 1.1 million tonnes of single-use plastics **EVERY YEAR!**

So, what can we do about this worldwide problem?

A lot of people are trying to come up with ways for the world to use less single-use plastic every day. I wonder if you have seen any of these ideas being used where you live?

Take a reusable shopping bag to the supermarket.	Take a reusable coffee cup to the coffee shop for takeaway coffee.	Put your lunch in a sealed container instead of wrapping it in plastic.

What other ways can you think of that people could do to reduce the amount of single-use plastics we use every day?

Here's a hint to get you started…Australians buy almost 15 BILLION plastic bottles every year, and most of them end up in landfill. The pictures around the boxes might also get you thinking.

Ways we could use less plastic:

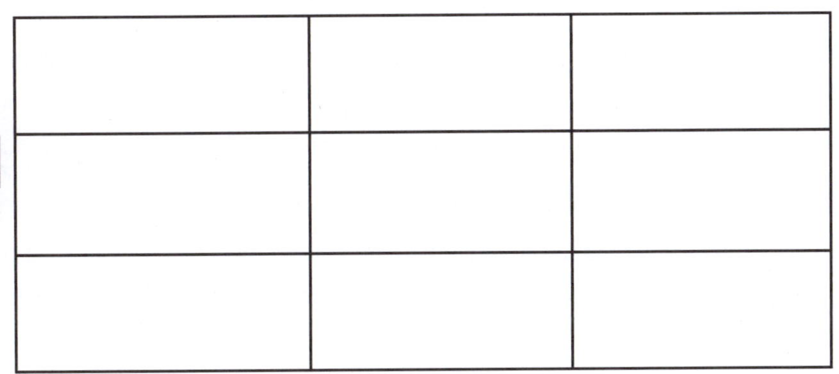

Use this QR code to access a video on this topic.

First Nations Perspective on Weapons

Traditionally, First Nations peoples needed to find whatever they required in the environment to construct tools, weapons, clothing and shelters to ensure their survival.

Knowledge of the useful parts of a plant, including the wood, bark, roots and leaves has been shared for millions of years. The natural properties of the various species of plants in each area means that different materials are selected for different purposes.

Dense woods are used to construct implements for striking and digging that need to be hard, heavy and durable.

Implements such as boomerangs are constructed from material that is strong but not heavy.

Fishing spears need to be made from a natural material that is lightweight, buoyant and flexible, so that the spear will float to the water surface to be retrieved after use.

Baur (multi-pronged fishing spears) are made from lightweight bamboo by the Meriam Peoples of the eastern Torres Strait Islands. However, in places like the mid north-west coast of New South Wales, fishing spears called **biguurr** are made from lightweight plants such as cottonwood and grass trees.

Spears used for hunting large animals, such as emu and kangaroo, needed a heavier, denser wood, so that it would have more of an impact when it hit the target. Paredarerme people of the Oyster Bay region of Tasmania constructed **perenna** (hunting spears) from tea trees.

Ooiritchanna were spears made by the Yankunytjatjara Peoples in the Musgrave Ranges. They were made from different woods, chosen for their natural properties. A light flexible wood from plants such as the wonga wonga vine is used for the spear shaft, and a hard, heavy material such as mulga is used to construct the spear head. The combination of materials meant that they had the best physical properties for a hunting spear - strength, flexibility and durability.

Page 3
1 producer, 2 decomposer, 3 consumer, 4 consumer

Page 4
1 mouse snake, 2 grasshopper mouse, 3 mouse snake

Page 5
1 dingo snake, 2 grasses, 3 frog mouse

Page 6
Out of the food chain consisting of grasses, antelopes and lions, the lion is at the top of the food chain.
1 impala, 2 grasses zebra and acacia tree giraffe, 3 grasses aardvark hyena

Page 7
1 predator, 2 carnivore, 3 herbivore, 4 prey, 5 competition, 6 food web, 7 omnivore, 8 producer, 9 food chain, 10 consumer

Page 9
Kookaburra (eats frog), moth (eats sedge, eaten by lizard, frog, dragonfly), lizard (eats ant, moth), frog (eats moth, ant, dragonfly, eaten by kookaburra), sedge (eaten by maned goose, moth, ant), dragonfly (eats moth, eaten by frog), maned goose (eats sedge), tadpole (eats algae), algae (eaten by tadpole)

Page 10
Answers should show an understanding of food chains and food webs. They should identify various producers, consumers, predators prey and show the complex interactions between them.

Page 11
1 plant roots, algae, fungi, bacteria, earthworms
2 Earthworms break down and recycle matter from dead plants.

Page 12
1 false, 2 true, 3 false

Page 13
3, 2, 1, 5, 4

Page 14
Insect consumer, earthworm decomposer, tomato plant producer
Answers should include understanding that consumers eat producers, and both consumers and producers are broken down by decomposers after death.

Page 15
1 (across) consume, 1 (down) castings, 2 microorganism, 3 recycle, 4 network, 5 decaying

Page 17
1 Earthworms ate the lettuce and mixed it into the soil, the other lettuce rotted and didn't move.
2 Earthworms mixed the layers, the other layers stayed in place.
3 Earthworms ate the lettuce and returned the nutrients to the soil. They burrowed and brough the nutrients in deep, missing them around.

Page 19
1 The temperature of the ocean varies.
2 The iguanas grow larger when the ocean gets colder, and shrink when it gets warmer.

Page 20
1 true, 2 true, 3 false

Page 21
1 Many animal species cannot survive without them.
2 Controlling the spread of paperbark trees.

Page 22
1 birds of the city, 2 birds of the forest, 3 64 km, 4 wetland birds

Page 23
1 native, 2 algae, 3 wildfire, 4 iguana, 5 ecosystem, 6 species, 7 affect, 8 extinct

Page 27
1 fresh water, 2 rain dams tanks etc, 3 not completely but there can be shortages

Page 28
pond, harbour, swamp, puddle, creek, sea

Page 29
waves large and breaking on shore, water level rises, river dries up

Page 30
It goes to a treatment plant and then back into the waterways.

Page 31
Take shorter showers, be careful watering the garden, wash the car on the lawn, don't hose down pavements, fix leaks in taps, don't play with water

Page 33
In water holes, in deep springs, in rock holes

Page 34
1 When water evaporates off the ground or from waterways.
2 It condenses, or forms tiny drops of water.

Page 37
1 cold, 2 snow

Page 38
1900-1950

Page 39
1 Yosemite valley, 2 the Great Lakes

Answers

Page 40

(top to bottom) glacier, moraines, basin

1 false, 2 true, 3 true, 4 true, 5 false

Page 41

1 very large, 2 retreating, 3 basins, 4 weight of the snow, 5 glaciers, 6 advancing, 7 ridges of dirt and gravel

Page 43

1 It scratched or tore the foil.

2 water and sand: a moraine

3 Glaciers can move rock, and these rocks can scratch and reshape the Earth's surface.

Page 44

Answers will vary, but may include: What is causing it? How can we slow it down? What effect will it have on ocean levels? What effect will it have on ocean currents? What effect will if have on wildlife? What effect will it have on the atmosphere?

Page 45

1 evaporates, 2 water cycle, 3 hydrosphere

Page 46

1 condensation, 2 condensation, 3 precipitation, 4 condensation

Page 47

(top left) precipitation, (top right) condensation, (bottom right) evaporation

Water returns to the sea during precipitation.

Page 48

1 hydrosphere, 2 humid, 3 precipitation, 4 condensation, 5 composed, 6 water cycle, 7 evaporates

Puzzle: biosphere

Page 49

Title: The Water Ccyle

a: evaporation (water turns from a liquid into a gas), b: condensation (water changes from a gas into a liquid); c: precipitation (water droplets fall to the ground)

Page 51

1 It evaporated, condensed on the plastic and then dripped into the cup.

2 Half a cup, about the same that we started with.

3 It made the condensed water gather over the cup so it could fall into it rather than onto the soil.

4 It could work but it would be slower, because the sunlight helps the water evaporate.

Page 52

Answers will vary but should show an understanding of the water cycle including evaporation, condensation and precipitation.

Page 53

1 false, 2 true, 3 false, 4 true

Page 54

1 groundwater, 2 3 per cent, 3 it is in the form of ice in ice caps and glaciers

Page 55

(top to bottom) well, aquifer, water table

Page 56

Pollution from agriculture and industry is seeping down through the ground into the aquifer. Seawater is flowing into the aquifer, making some of it too salty. The groundwater has fallen to a low level in the well, meaning there is not much to access.

Page 57

1 pollution, 2 irrigation, 3 conservation, 4 dissolve, 5 sustain, aquifer, 7 water table, 8 groundwater

Page 59

water on white sand: it soaks into the sand

water on clay: it rolls of the clay onto the sand

water on rocks: it fills up the space between the rocks and makes a little lake

food colouring in cup: it spreads through the water

1 clay as it is does not have space in it for the water to fill, 2 the water poured over the rocks, 3 the top of the water in the cup, 4 pollution soaks through surface water, rock and soil to reach the groundwater

Page 60

1 bathroom, 2 watering garden swimming pools

shorter showers, turn off tap while brushing teeth, water-saving devices, recycle grey water for garden, do only full loads of laundry

Page 61

(left to right) push, pull, pull

Page 62

1 tyres and brakes, 2 sled

Page 63

1 change the shape of things, 2 direction

Page 64

1 unbalanced, 2 unbalanced (arrows of different lengths), 3 balanced, 4 balanced

Page 65

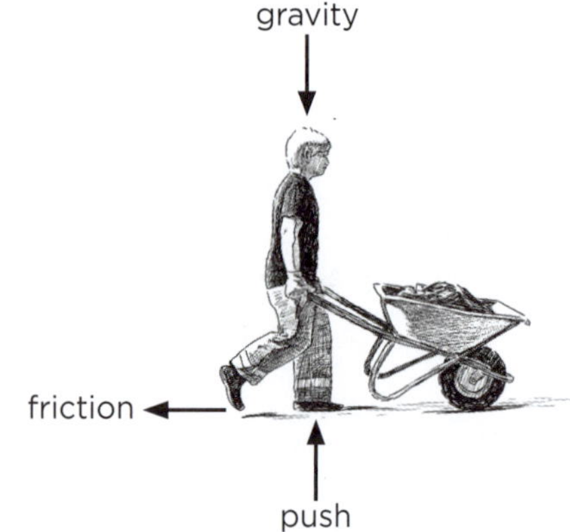

TARGETING SCIENCE YEAR 4 © PASCAL PRESS ISBN: 978125726534

Page 66
marble, fishing sinker, rock, metal coil, fork, blob of clay

Page 67
contact forces: hitting a tennis ball, blowing a feather, rubbing your hands together, a plane's propellers, catching a frisbee
non-contact forces: a rock dropping to the ground, electricity, a magnet pushing another, the moon in space

Page 69
1 force, 2 force, 3 equal, 4 slow things down, 5 in motion, 6 smooth, 7 where it goes, 8 unequal, 9 speed, 10 unseen
Real, because of the effect it has on objects that we can see.

Page 70
1 By pulling back harder on the balloon.
2 The greater the force, the longer the distance travelled.

Page 71
smooth foil

Page 72
1 false, 2 true, 3 true, 4 false

Page 73
1 potential energy, 2 the energy of motion, 3 Mason's snowball

Page 74
1 sound, 2 light, 3 electric current

Page 75
1 heating, 2 electronics, lighting and other appliances, 3 space heating, electronics and other appliances, water heating

Page 76
2 It is still an egg white, energy only changed the way it looked.

Page 78
The bat is pulled back, storing potential energy. When it hits the ball, it transfers to the ball in the form of kinetic energy.

Page 79
1 gravity pulls things towards the Earth's centre, 2 Newton

Page 80
1 earth, 2 earth

Page 81
You would weigh less on the moon because it has a smaller mass than the Earth, and therefore a lower gravity. Weight measures the pull of gravity.

Page 82
4, 5, 1, 2, 3, gravity pulls it down

Page 83
1 weight, 2 Newton, 3 weightless, 4 matter, 5 gravity, 6 attract, 7 astronaut, 8 scientist

Page 87
Playing sport, walking, dancing, jumping, falling over

Page 88
1 force, 2 centre, 3 sphere

Page 89
1 true, 2 true, 3 false, 4 true

Page 90
1 downwards, 2 all at the same time, 3 downwards, 4 the swing, 5 a force

Page 91
1 orbit, 2 sphere core, 3 air resistance, 4 exert mass, 5 gravity, 6 encounter, puzzle: Mercury

Page 93
1 ball rock cotton, 2 feather, 3 the feather because it floated and took longer to fall, 4 they would all fall at the same speed as there is no air resistance

Page 94
Answers will vary, but should show an understanding of how gravity works and the effects of gravity.

Page 95
1 coin, glass, plastic bricks, 2 paperclip, safety pin, tin can - can test them by seeing if a magnet sticks to them

Page 96
1 B, 2 A

Page 97
1 A, 2 A

Page 98
1 repel: alike poles repel, 2 repel: alike poles repel, 3 attract: opposite poles attract

Page 99
1 field, 2 iron, 3 repel, 4 poles, 5 wood.
Riddle: field

Page 104
Soft things: rubber, rubber bands, blue tack; **Wooden things:** pencil, ruler; **Flexible things:** notebook, rubber, rubber bands, blue tack; **Round things:** magnifying glass, ball of rubber bands; **Hard things:** pencil, scissors, ruler, magnifying glass; **Opaque things:** notebook, pencil, rubber, scissors, ruler, rubber bands, blue tack; **Stretchy things:** rubber bands, blue tack; **Smooth things:** notebook, pencil, rubber, scissors, ruler, magnifying glass, blue tack

Answers

Page 105

Sticky tape is sticky so things will stick and the holder is rigid to keep tape in place.

Scissors are smooth and rigid for ease of use with sharp blades for cutting.

Texta is smooth and rigid to hold the wet smooth ink tip and stop it drying out.

Skipping rope has soft but strong handles with a thin rope for skipping.

Page 107

L	S	G
S	G	L
S	L	G
S	L	S
S	L	G

Page 109

P	P	C
P	C	P
P	P	P
C	C	C

Page 110

1 The water was gone and the salt was left behind, wherever the drips were.

3 The water had evaporated but the salt did not.

4 Physical. Because the substances could go back to their original forms.

Page 112

2 It soaked up the most water

3 So that it does not let liquids through when used for cooking

5

A	W	A	W
A	A	W	A
W	W	W	A
W	W	A	A

Page 115

1 to feed wild birds, 2 they hold and dispense bird seeds, 3 they are made of different materials, 4 what sort of birds you have and what your climate is like